Social Computing

and

Virtual Communities

Edited by

Panayiotis Zaphiris *and* Chee Siang Ang

CRC Press
Taylor & Francis Group
Boca Raton London New York

CRC Press is an imprint of the
Taylor & Francis Group an **informa** business

A CHAPMAN & HALL BOOK

Chapman & Hall/CRC
Taylor & Francis Group
6000 Broken Sound Parkway NW, Suite 300
Boca Raton, FL 33487-2742

First issued in paperback 2017

ISBN 13: 978-1-4200-9042-0 (hbk)
ISBN 13: 978-1-138-11613-9 (pbk)

Library of Congress Cataloging-in-Publication Data

Social computing and virtual communities / Panayiotis Zaphiris, and Chee Siang Ang, editors.
    p. cm.
  Includes bibliographical references and index.
  ISBN 978-1-4200-9042-0 (alk. paper)
    1. Human-computer interaction. 2. Internet--Social aspects. 3. Computer networks--Social aspects. I. Zaphiris, Panayiotis. II. Ang, Chee Siang.

QA76.9.C66S619 2010
004.01'9--dc22
                2009021896

Visit the Taylor & Francis Web site at
http://www.taylorandfrancis.com

and the CRC Press Web site at
http://www.crcpress.com

Contents

Preface

Advances in computer technology, particularly the development of the Internet, have affected virtually every aspect of society. Unsurprisingly, research and development in the area of social computing, as well as that of virtual communities, has emerged and is increasingly gaining importance. This book brings together contributions from international experts from various fields and strives to explore the social use of computers. Social computing and virtual communities not only improve productivity and learning outcomes via collaborative work and learning but also augment human relationships by providing various means of communication that nullify the geographical barriers existing among us. In addition, they transform the platform of digital entertainment, infusing a social dimension into the already popular use of the Internet. We are witnessing the formation of persistent and long-term communities in a virtual setting, where people congregate to socialize, make friends, give support, etc. Virtual communities continue to grow in size, nature of use, and technological aspects, such as the emergence of three-dimensional (3D) virtual environments.

Due to the escalating use of computers to mediate various social aspects of human activity, some conceptual frameworks have been developed to understand computer-mediated social activities. It is commonly agreed that researching, designing, and evaluating virtual communities require a multidisciplinary approach that goes beyond methods used for designing and evaluating other interactive systems. Furthermore, it is also crucial to research these issues through naturalistic approaches in which users' activity is investigated in context. Therefore, this book aims at presenting the topic of social computing and virtual communities from the user's perspective in various domains of use through a multidisciplinary lens.

One of the most important characteristics of the Internet is the opportunities it offers for human–human communication through computer networks. As Metcalfe (1992) points out, communication is the Internet's most important asset. E-mail is just one of the many modes of communication that can occur through the use of computers. Jones (1995) points out that through communication services like the Internet, Usenet, and bulletin boards, online communication has for many people supplanted postal services, telephones, and even fax machines. These applications in which the computer is used to mediate communication are called computer-mediated communication (CMC).

Studies of CMC view this process from different interdisciplinary theoretical perspectives (social, cognitive/psychological, linguistic, cultural, technical, and political) and often draw from fields as diverse as human communication, rhetoric and composition, media studies, human–computer interaction, journalism, telecommunications, computer science, technical communication, and information studies.

Online communities emerge through the use of CMC applications. The term "online community" is multidisciplinary in nature, meaning different things to different people, and is slippery to define (Preece, 2000). In defining CMC, Rheingold (1993) states that

> [Online] communities are social aggregations that emerge from the Net when enough people carry on those public discussions long enough, with sufficient human feeling, to form webs of personal relationships in cyberspace. (p. 5)

Online communities are also often referred to as cyber societies, cyber communities, Web groups, virtual communities, Web communities, virtual social networks, and e-communities among several others.

Cyberspace is the new frontier in social relationships, and people are using the Internet to obtain colleagues, find lovers, and make friends—as well as enemies (Suler, 2004). As Korzeny pointed out, as early as 1978, online communities were formed around interests, and not physical proximity (Korzeny, 1978). In general, what brings people together in an online community is common interests such as hobbies, ethnicity, education, and beliefs. As Wallace (1999) points out, meeting in online communities eliminates prejudging based on someone's appearance, and thus people with similar attitudes and ideas are attracted to each other.

These issues are addressed in this book through a collection of chapters contributed by the leading experts in these areas. The book is divided into three parts—I: Theories and Methods; II: Application Areas; and III: Types of Online Social Environments.

Part I consists of three chapters, which cover theoretical issues and methodological concerns in studying and research virtual communities.

In Chapter 1, Charalambos Vrasidas and George Veletsianos explore and examine the theoretical aspects of social computing and virtual communities from the perspective of cognition, constructivism, and community of practice. Several theories regarding online social interaction are outlined and a case study is described to demonstrate how theories are put into practice. This chapter serves as a good introduction to other topics in this area covered by this book.

Moving from theories, in Chapter 2, Gindo Tampubolon draws our attention to an increasingly popular and important method in analyzing social interaction either online or offline, social network analysis (SNA). This technique is explained tactfully in plain language with plenty of real-life examples and does not assume any mathematical background to understand.

In Chapter 3, Paul Cairns and Mark Blythe provide a general overview of methodologies for analyzing and evaluating virtual communities. Both qualitative and quantitative methods are discussed in relation to analyzing data gathered from the online setting. A case study is then presented to demonstrate how various methods can be applied to analyze online activities on YouTube™, a video-sharing Web site.

Part II has five chapters that include a whole range of application domains of social computing such as learning, healthcare, etc.

Perhaps the most prominent use of social computing recently is e-learning as more and more higher education institutions have adopted some forms of social computing technologies. In Chapter 4, Andrew Laghos provides an in-depth investigation into e-learning communities using methods covered in Chapters 2 and 3 with a particular emphasis on the formation of social networks of learners. A framework of analyzing e-learning communities is also presented.

Chapter 5 examines an interesting domain, namely, healthcare communities. Laura Slaughter, Aksel Tjora, and Anne-Grete Sandaunet highlight several important issues in a community motivated by physical pain, illness, and burdensome symptoms. An overview is outlined, focusing on various types of healthcare communities, current research questions, and methodologies used to study these communities. Two case studies are also described to provide practical examples.

A complement to the previous chapter, Chapter 6 delves into issues of online support and empathy, virtual communities for people to exchange emotional support. In this chapter, Ulrike Pfeil illustrates the unique characteristics of such communities. Advantages and challenges of creating a virtual space for social support are described. This chapter also highlights method triangulation in analyzing virtual communities, emphasizing the importance of validity issues using multiple methods.

In Chapter 7, Masahiro Hamasaki, Kouchiro Eto, Sri Kurniawan, Tom Hope, Hideaki Takeda, and Takuichi Nishimura underline the heated issue of intellectual property (IP) in an online setting by presenting a case study on content sharing. Privacy and IP issues have recently caught the attention of researchers and policy makers alike. This chapter explores in depth an interesting online phenomenon—user-generated content (UGC)—and its implication on these issues with an aim to enhance our understanding to develop not only technological but also social and legal solutions.

Chapter 8 explores yet another intriguing topic, online trust. Paula Bialski and Dominik Batorski look into what is known as a hospitality network in an online setting. Individuals use this network as an online space to seek accommodation in the home of another member in the network. In such a virtual community, trust is central, and the question of trustworthiness is made explicit in the design of such a social system.

Part III, consisting of four chapters, closely examines different types of social computing and social technology.

Chapter 9 introduces a new type of social computing and virtual community, in which users immerse themselves in a 3D virtual environment beautifully rendered in cutting-edge computer graphics. This type of 3D world raises several issues, as Diane Carr and Martin Oliver note in this chapter, such as self-image projection with avatars. This chapter emphasizes the issue of immersion and its relation to using a 3D virtual world, namely, *Second Life*™ in learning.

Chapter 10 expands the discussion of 3D virtual worlds and focuses on a specific genre, MMORPG (massively multiplayer online role-playing games). David White in this chapter provides an ethnographic account of the "virtual life cycle" of the users in this social gaming environment, from becoming a resident and gaining social skills to forming close relationships and a guild community.

Chapter 11 presents a very unique type of community consisting of older people from the Chinese culture. Bo Xie explores the perceptions of the Internet and virtual communities among older Chinese people. It is found that unlike many other types of communities, which often see the Internet as a new medium that facilitates new kinds of lifestyles, the Chinese elderly perceive virtual communities as a place where old beliefs and traditional norms can be preserved in contemporary China.

Finally, in Chapter 12, Mike Thelwall and David Stuart conclude by describing the most recent development of social computing, social networking sites (SNS). Unlike traditional virtual communities, SNS consists of social networks of people, which expand quickly. These virtual social networks are often connected to offline networks, triggering various issues, such as cultural differences, identity, etc.

Altogether, we believe that this collection will benefit anyone who is interested in this new and exciting area that is growing and expanding rapidly. The book not only provides case studies of different domains of virtual communities and different types of social technologies but also emphasizes theoretical and methodological aspects of researching and analyzing such communities. Books of this kind are uncommon, and we hope this collection, drawing from international experts of different areas, will cast some light on the topic of social computing and virtual community. Above all, we hope you enjoy reading the book.

References

Jones, S. (1995). Computer-mediated communication and community: Introduction. *Computer-Mediated Communication Magazine*, 2:3.

Korzeny, F. (1978). A theory of electronic propinquity: Mediated communication in organizations. *Communication Research*, 5:3–23.

Metcalfe, B. (1992). Internet fogies to reminisce and argue at Interop Conference. *InfoWorld*, p. 45.

Preece, J. (2000). *Online Communities: Designing Usability, Supporting Sociability*. Chichester, UK: John Wiley and Sons.

Rheingold, H. (1993). *The Virtual Community: Homesteading on the Electronic Frontier*. Reading, PA: Addison-Wesley.

Suler, J. (2004). The final showdown between in-person and cyberspace relationships. Retrieved March 3, 2009 from http://www-usr.rider.edu/~suler/psycyber/showdown.html.

Wallace, P. (1999). *The Psychology of the Internet*. Cambridge, UK: Cambridge University Press.

About the Editors

Panayiotis Zaphiris is an associate professor at the Department of Multimedia and Graphic Arts of Cyprus University of Technology (CUT). Before moving to CUT, he was a reader at the Centre for Human–Computer Interaction Design, School of Informatics of City University London. He earned his Ph.D. in Human Computer Interaction (HCI) from Wayne State University in the United States. His research interests lie in HCI with an emphasis on inclusive design and social aspects of computing. He is especially interested in HCI issues related to the elderly and people with disabilities. He is also interested in Internet-related research (Web usability, mathematical modeling of browsing behavior in hierarchical online information systems, online communities, e-learning, Web-based digital libraries, and, finally, social network analysis of online human-to-human interactions).

Chee Siang Ang is a lecturer at the School of Engineering and Digital Art, University of Kent. Before that, he was a research fellow at the Centre for Human–Computer Interaction Design, School of Informatics, City University London, where he completed his Ph.D. in the area of social gaming. He holds a Master's degree in information technology from the Multimedia University of Malaysia, and he obtained his BSc. in computing from the Technology University of Malaysia. His research interests center on aspects of human–computer interaction, particularly usability and sociability, of computer games and three-dimensional virtual worlds. He is also interested in social network analysis in the virtual context, digital media (such as interactive narratives and simulations), and the serious use of digital media in, for example, learning and teaching.

Contributors

Dominik Batorski
Assistant Professor
University of Warsaw
Warsaw, Poland

Paula Bialski
Research Student
Lancaster University
Lancaster, United Kingdom

Mark Blythe
Senior Research Fellow
University of York
York, United Kingdom

Paul Cairns
Senior Lecturer
University of York
York, United Kingdom

Diane Carr
Lecturer
Institute of Education
University of London
London, United Kingdom

Kouchiro Eto
Researcher
National Institute of Advanced
    Industrial Science and Technology
Tokyo, Japan

Masahiro Hamasaki
Researcher
National Institute of Advanced
    Industrial Science and Technology
Tokyo, Japan

Tom Hope
Researcher
National Institute of Advanced
    Industrial Science and Technology
Tokyo, Japan

Sri Kurniawan
Assistant Professor
University of California, Santa Cruz
Santa Cruz, California, USA

Andrew Laghos
Lecturer
Cyprus University of Technology
Lemesos, Cyprus

Takuichi Nishimura
Researcher
National Institute of Advanced
    Industrial Science and Technology
Tokyo, Japan

Martin Oliver
Reader
Institute of Education
University of London
London, United Kingdom

Ulrike Pfeil
Research Student
City University London
London, United Kingdom

Anne-Grete Sandaunet
Postdoctoral Researcher
University of Tromsø
Tromsø, Norway

Laura Slaughter
Research Scientist
Norwegian University of Science and
    Technology
and
Rikshospitalet University Hospital
Oslo, Norway

David Stuart
Research Fellow
University of Wolverhampton
Wolverhampton, United Kingdom

Hideaki Takeda
Researcher
National Institute of Advanced
    Industrial Science and Technology
Tokyo, Japan

Gindo Tampubolon
Research Fellow
University of Manchester
Manchester, United Kingdom

Mike Thelwall
Professor
University of Wolverhampton
Wolverhampton, United Kingdom

Aksel Tjora
Professor
Norwegian University of Science and
    Technology
Trondheim, Norway

George Veletsianos
Lecturer
University of Manchester
Manchester, United Kingdom

Charalambos Vrasidas
Associate Professor
University of Nicosia
Nicosia, Cyprus

David White
Manager of Technology Assisted
    Lifelong Learning
University of Oxford
Oxford, United Kingdom

Bo Xie
Assistant Professor
University of Maryland
College Park, Maryland, USA

THEORIES AND METHODS

Chapter 1

Theoretical Foundations of Social Computing and Virtual Communities

Charalambos Vrasidas and George Veletsianos

Contents

1.1 Introduction

Neglecting the sociopsychological dimension of technology-mediated interactions impedes engagement with virtual environments. Successful research and practice depends on the theoretical foundations of social computing and virtual communities. Our purpose in this chapter is to examine the theoretical foundations of the social nature of computer use and virtual communities. We begin by examining the social nature of human activity, the characteristics of communities and social computing, and we propose an interactional theoretical framework within which the fundamental process and unit of analysis is "interaction." We will discuss factors that enhance technology-mediated social environments and examine theories that may promote successful engagement with such environments. Theoretical underpinnings surveyed include the social nature of human activity and learning, the notion of affordances, situated cognition, computer-supported collaborative learning (CSCL), and social presence theory. We also discuss emerging technologies and how they facilitate social activity, and we present an extensive practical example delineating how theory links to practice.

1.2 The Social Nature of Human Activity

Human activity is by nature social. Our work is informed by *symbolic interactionism* ideas, according to which there are three basic premises (Blumer, 1969). First, human beings act upon the world on the basis of the meanings that the world has for them. Second, the meaning of the world is socially constructed through one's interactions with members of the community. Third, the meaning of the world is processed again through interpretation. Traditional approaches to research tend to ignore the importance of meaning, interaction, and interpretation of the actors in shaping human behavior. Rather, traditional research has pushed aside meaning

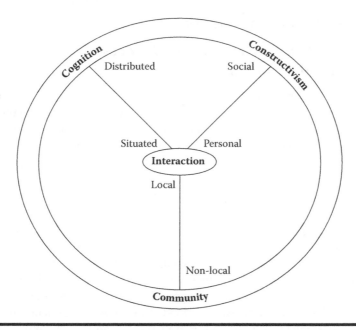

Figure 1.1   Interaction as a central component of human activity.

as unimportant, attempting to explain behavior as a result of a set of factors. An action is the observable behavior plus the meaning attached to it by the actor. In Blumer's (1969) words, in traditional research approaches, "meaning is either taken for granted and thus pushed aside as unimportant or it is regarded as a mere neutral link between the factors responsible for human behavior and this behavior as the product of such factors" (p. 2). Therefore, examining the socially constructed meaning is important in understanding human activity.

Using a symbolic interactionist framework, interaction is defined as the reciprocal actions of two or more actors within a given context (Vrasidas and Glass, 2002). Interaction is an ongoing process that resides in a context and also creates context. Context is crucial in examining human activity in a sociotechnological environment. Context is provided by the history of the situation, past interaction sequences, and the anticipation of future interaction sequences. There is a reflexive relationship between context and interaction that prevents us from isolating the two. To examine human activity, it is important to carefully study the context and the moment-to-moment events that lead to further interaction among people. Human beings are active, meaning-making living organisms. Social interaction is an ongoing process that shapes human conduct as actors fit their actions with one another and form a continuous flow of interaction. Participants have intentions that influence interaction.

1.3 Theoretical Framework

1.3.1 Interaction: The Fundamental Process of Social Computing

During the past fifteen years, members of the CARDET (Centre for the Advancement of Research and Development in Educational Technology) team have conducted a series of studies to examine the challenges and possibilities of online environments (e.g., Vrasidas and Glass, 2005; Vrasidas et al., 2009; Zembylas et al., 2002). The results of those studies led to the construction and refinement of a conceptual framework that draws from three interrelated areas: social constructivism (Brown et al., 1989; Lave and Wenger, 1991; Vygotsky, 1978), situated and distributed cognition (Brown et al., 1989; Salomon, 1993), and communities of practice (Wenger, 1998). In the sections below, we briefly discuss each of these components, focusing on their interconnection and implications for the design of online environments. A fundamental idea that is embedded in these three components is *interaction*. This framework places interaction at the center of human activity (Vrasidas and Zembylas, 2004).

In this framework, interaction takes multiple forms in the three-dimensional space (Figure 1.1). Each dimension is divided into two components; these components are positioned in a continuum and there is no clear boundary between them. Theoretically, online environments and virtual communication can be situated somewhere along the continua of these three dimensions. It is important to note that interaction as a process is constantly changing and adjusting to the needs of the community and, although it already has a predefined structure, is constantly emergent, negotiated, and renewed as activities and contexts dictate.

1.3.2 Personal and Social Constructivism

Over the past two decades, constructivism has been embraced in a variety of fields. Although there are various kinds of constructivism, a fundamental assumption is that knowledge does not exist independent of the learner, i.e., knowledge is constructed. Several philosophers and educators are associated with constructivism (e.g., von Glasersfeld, 1989; Vygotsky, 1978). The two most prominent schools of thought within constructivism are *personal* constructivism and *social* or sociocultural constructivism (Cobb, 1994). Their major difference has to do with the locus of knowledge construction (Vrasidas, 2000). For personal constructivism, knowledge is constructed in the head of the learner while he/she is reorganizing his/her experiences and cognitive structures (Piaget, 1970; Von Glasersfeld, 1989). For social constructivism, knowledge is constructed in "communities of practice" through social interaction (Brown et al., 1989; Lave and Wenger, 1991; Vygotsky, 1978). Our view is that the two approaches cannot be separated because they complement each other. Knowledge is both constructed through social interaction and in the learner's mind; there is no clear boundary between these two processes.

1.3.3 Situated and Distributed Cognition

The ideas of situated and distributed cognition are useful in helping us concep-tualize the ways that knowledge and cognition can be both local and distributed. One of the early thinkers on the ideas of distributed cognition is Salomon (1993), who argued that "with a growing acceptance of a constructivist view of human cognitions comes serious examination of the possibility that cognitions are situated and distributed rather than decontextualized tools and products of mind" (p. xiv). Brown et al. (1989) proposed the concept of *situated cognition* and argued that activities during which knowledge is constructed constitute an integral part of that knowledge. The situation in which knowledge is constructed is an integral part of the learning process, i.e., knowledge and cognition are situated. Learning becomes a process of enculturation as learners are immersed in real-life situation and act as experts (Lave and Wenger, 1991). As Lave and Wenger (1991) point out, "Learning is a process that takes place in a participation framework, not in an individual mind. This means, among other things, that it is mediated by the differences of perspective among coparticipants. It is the community, or at least those participat-ing in the learning context, who "learns" under this definition. Learning is, as it were, distributed among coparticipants, not a one person act" (p. 15). Therefore, our framework is based on the notion that knowledge and cognition are *distrib-uted* among online participants and their physical and sociopolitical and historical worlds (Pea, 1993; Vygotsky, 1978). The knowledge that we have and need is not all in our head, but to a large extent resides in the world—in artifacts of all kinds and in the minds of other people. Distributed cognition contributes to our theoretical framework in that it enriches understanding about what individuals can achieve and how artifacts, tools, and sociotechnical environments can be designed and evaluated to empower individuals and their activity (Vrasidas and Zembylas, 2004).

1.3.4 Local and Nonlocal Communities of Practice

The relationship between technologies and communities is complex. On the one hand, changing technologies of communication and cooperation have facilitated the participation in *nonlocal* communities. On the other hand, the role of place remains important, as the *local* environment is a significant place of organizing and coordi-nating social life. Wenger (1999) has described four dualities that characterize com-munities: participatory-reification, designed-emergent, identification-negotiation, and local-global. For our work, we address the last one, which we renamed "local versus nonlocal," because we have criticized elsewhere the notion of global com-munities as something highly problematic (see Vrasidas and Zembylas, 2003; Zembylas et al., 2002; Vrasidas et al., 2009). We have argued, for example, that common metaphors such as the *global village* may not always work when applied to experiences of marginalized people in online communications and thus, regard-less of physical access, the online environment can be "exclusionary." Some aspects regarding the other three dualities are addressed within our discussion regarding distributed-situated cognitions and individual and social constructivism.

In the past two decades, there is increasing interest in constructing e-learning spaces to support communities of practice (Kim, 2000; Schlanger and Fusco, 2002; Schwen and Hara, 2003; Vrasidas and Glass, 2005). Listservs, bulletin boards, course management systems, and e-learning environments can offer alternative methods for the development of communities. What characterizes communities of practice is a shared commitment for a particular practice that creates an interactional network enabling and promoting knowledge sharing and professional development (Hoadley and Pea, 2002; Wenger, 1999; Wenger et al., 2002).

Wenger (1998) proposed the four dimensions of learning as they are worked out in action from within a community of practice framework: learning as doing (practice), learning as becoming (identity), learning as experience (meaning), and learning as belonging (community.) Communities of practice are groups of individuals bound by what they do together—e.g., from engaging in informal discussions to solving problems—and by what they have learned through their mutual engagement in these activities. Rules of engagement within a community of practice are constantly renegotiated although there is a shared repertoire of communal activities, routines, discourses, and so on that members have developed over time. Thus, communities of practice have been theorized as sites of mutual learning and as important contributors to the success of knowledge-dependent organizations.

In online environments, communities are growing and are developing new ways of using information and communication technologies. The interest in online communities grows day by day, and corporations and education institutions alike are utilizing the power of online community building for lifelong learning and continuing education. Thus, understanding the challenges for the design of online communities of practice is important. For example, in our studies we have used a variety of means to understand the needs of an online community of practice, such as quantitative studies of communication patterns (i.e., social network analysis) and resource usage, and qualitative (i.e., ethnographic) studies of how individuals and groups conduct their work and their communications and how they express their needs.

1.3.5 Activity Theory

The roots of activity theory (AT) can be traced back to the 1930s and the works of Soviet scientists Rubinshtein, Leont'ev, and Vygotsky. Although much of the Russian literature has yet to be translated, numerous scientific communities across the world have sought to understand and promote AT. Most notably, Cole and Engeström have both been significant contributors in researchers' understanding of AT in English-speaking nations (e.g., Cole and Engeström, 1993; Engeström, 1987, 1999). Our writing here is based on their work and more specifically on Cole and Engeström (1993).

AT views social practice as central to learning. Unique to AT, however, is the idea that activity is seen through activity systems, the smallest units of activity, that are being regulated by rules, outcomes, divisions of labor, and community, and mediated by tools (e.g., language and technologies.) In other words, AT provides a lens through which we can observe how individuals use tools to shape social contexts and how tools used within those systems shape the individuals. An important facet of AT is the interdependence rather than independence of activity elements: The subject and the object are mediated by tools, and neither can be understood by themselves. In the same way, rules, outcomes, divisions of labor, and community influence the activity, which itself is embedded within a unique sociocultural and historical context. The complexity of human activity and its interrelated components is illustrated in Figure 1.2 (Engeström, 1987).

To further clarify the notion of activity systems and provide the reader with a contemporary example, consider the following activity system. Let's imagine a collaborative blog, co-owned by blog contributors, dedicated to informing readers about news and events related to technology gadgets such as MP3 players, digital cameras, and mobile phones. In this system, the activity system components might be envisioned to be as follows:

Participant: Blog author
Mediating artifacts: Digital tools and language (e.g., Wordpress blogging platform)
Rules: Gather gadget related news from a variety of sources; post breaking and informative news as often as possible; posts should be interesting, entertaining, and truthful

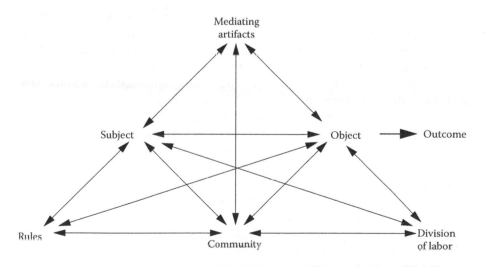

Figure 1.2   A model of the structure of human activity.

Outcomes: Increasing blog traffic; improved revenues (from the blog author's perspective, fulfilling a partnership requirement, earning industry recognition)

Division of labor: Blog authors as active owners rather than employers or employees; democratic participation; shared content ownership

Community: Industry partners; blog readers (from the blog author's perspective, building a relationship around the blog's topic with blog readers.)

It is also important to note that, even within a single activity, multiple activity systems might coexist. For instance, in the fictitious scenario presented above, a blog contributor might consider critical gadget reviews rather than breaking news to be the rules that mediate her activity within the aforementioned system. Doering et al. (2008) illustrate the notion of multiple activity systems when they describe the intended activity system that they planned for their students before a classroom assignment. It was quickly discovered, however, that intentions and outcomes were not in harmony: "[The] single assignment—which we considered to be part of one activity system—consisted of three overlapping activity systems that complicated the assignment" (p. 49). A number of unforeseen factors led to students developing their own activity system, with its own rules, outcomes, and community norms. In line with communities of practice theory (Wenger, 1998), subjects, acting in their best interest and, by implication, in the best interests of the organization to which they belong, structured their activity around practices that were seen as noteworthy.

1.3.6 Affordances

Affordance is a notion borrowed from Gibson's (1977, 1979) work. An affordance of a thing refers not only to the properties of the thing that allow it to be used in certain ways but also to the perceived properties of the thing (Norman, 1988). For example, a chair affords sitting; but it also affords standing on and balancing upon. Interactive television affords seeing and hearing individuals from miles away and interacting with them in real time. Telecommunications technologies afford synchronous and asynchronous interaction among multiple users. Computer conferencing and the Internet afford the design of social computing environments that support the development of communities of inquiry, collaboration, negotiation, and problem solving within authentic contexts (Vrasidas, 2000). Technologies in wide use at the time some distance education theories were proposed could not afford learner-learner interaction at a frequent rate and on a large scale (Vrasidas and Glass, 2002). The affordances of new telecommunications technologies and personal computers prompt us to revisit fundamental assumptions about distance education pedagogy. What we know about pedagogy requires that the technological affordances of the future be driven in part by the needs of education, even as invention in the past was driven by the exigencies of business.

Information communication technologies that mediate interaction are important within this framework. Technologies used in distance education, for example, include books and other print material, broadcast television, video, audio, telephones, and computers. Various technologies allow different cues to be communicated and different communication strategies. Print material used in correspondence education was limited in the kinds of interactions allowed and was more appropriate for one-way communication. Affordances of telecommunication technologies (e.g., computer conferencing, interactive video) allow two-way communication and interaction.

1.3.7 Social Presence Theory

Short et al. (1976) coined the term "social presence" and originally defined it "as a quality of the medium itself" (p. 65). According to these authors, different media engender different degrees of social presence. Social presence is the degree to which a medium allows the user to feel socially present in a situation that is mediated via technology. It is hypothesized that the more cues transmitted by a medium, the more social presence characterizes the medium. Therefore, media such as interactive video and multimedia computer conferencing carry with them the potential for higher degrees of social presence than do text-based computer-conferencing systems.

Several studies were conducted to examine the construct of social presence. Gunawardena (1995) examined whether social presence is a characteristic of the medium itself or whether it is an attribute of the user's perception of the medium. The findings of her study showed that social presence can be promoted by employing strategies that encourage interaction. Her research showed that the lack of audiovisual cues led online users to invent other means for compensating for the lack of those cues. These findings are in agreement with other studies that show that the temporal, spatial, and contextual limitations of online interaction lead to the development of strategies that compensate for the lack of visual and aural cues present in face-to-face interaction (Gunawardena and Zittle, 1997; Rezabek and Cochenour, 1998). Such strategies include the use of capitalization, abbreviating the message, and the use of emoticons.

Still debated is whether heightened social presence is always to be sought after. Some individuals may find the "face-to-face ideal" interaction too anxiety provoking and may function better in an environment that puts more distance between participants (Vrasidas and Glass, 2002). Although it is often assumed that immediate interactions are preferred, Burbules (2000) argued that "it is a myth to imagine that more immediate interactions are always the most honest, open, and intimate ones" (p. 329). For example, some learners feel more comfortable and less emotionally intimidated interacting with teachers and peers by not being physically present (Burbules and Callister, 2000). A more definitive example comes from the computer science literature related to awareness with respect to videoconferencing and instant messaging applications. Although in the early 1990s it was assumed that videoconferencing and instant messaging systems should approximate physical colocation, it was quickly decided that heightened awareness was neither necessary at all times and costs nor, most importantly, desirable (Edigo, 1988; Olson and Olson, 2008).

1.3.8 Summary

To summarize the three components of our framework—individual and social constructivism, situated and distributed cognition, and local and nonlocal communities of practice—we focus on three concepts that provide the foundations for the interrelations among these components: interaction, meaning, and enculturation. If there is a common core to be claimed for this framework, it might be this: a concern for the mediated nature of cognition and knowledge, i.e., the interactional view frames action and meaning-making as something that is mediated. This renders knowledge and cognition as participatory, distributed, and culturally constructed. In other words, we interact with the world through mobilizing cultural resources into our actions.

We argue that adoption of this perspective has implications for the study of social computing because it places at the center of study interaction, which is the fundamental process in social computing and community participation. Our discussions have demonstrated how the use of technology implicitly or explicitly endorses this interactional view of social computing. In the following sections, we discuss in more detail some of the theoretical issues that can inform the study of social computing.

1.4 Technology and Social Activity

1.4.1 Computer-Supported Collaborative Work/Learning

A popular term used to describe technologies that enabled collaboration in the 1970s and 1980s was "groupware." Such technologies may include items such as online social networking environments (e.g., MySpace), social computing development tools (e.g., wikis), document sharing systems (e.g., email), shared computing systems (e.g., public displays), and other collaborative environments (e.g., teleconferencing). Although we cover contemporary social computing tools and practices later in this chapter, it is worth noting that computer-supported collaboration has been an issue of conversation and practice since at least the 1980s (Grudin, 1994). Computer-Supported Collaborative Work/Learning (CSCW/L) refer to the study of how people work or learn together using technology systems and applications. Although the focus on CSCW/CSCL appears to be on facilitating group processes and interaction such that work or learning is attained more effectively and efficiently, it is also evident that there exists no uniform consensus on what the study of CSCW/L entails or purports to do (Bannon and Schmidt, 1991).

The use of technological innovations to harness collaboration has been an important part of social computing; yet, it is important to recognize that issues unrelated to technology influence the success of CSCW and CSCL. Such issues may range from work/learning culture and purpose of using a collaborative tool to pedagogical soundness and perceived benefit of using a tool. Recently, the importance of activity theory in understanding and contextualizing computer-supported work and learning has been highlighted (e.g., The *Journal of Computer Supported Cooperative Work* hosted a special issue—2002, Volume 11, Numbers 1-2—on activity theory and the practice of design).

1.4.2 Virtual Social Networking and Web 2.0

The newly born World Wide Web (WWW) of the early 1990s was characterized by content publishing, communication, and collaboration. Although the individuals publishing content in the early 1990s were those who were comfortable with computing, the content creators of the early 2000s are not necessarily computing experts in the sense of writing code and tinkering with software. The ten years since the birth of the web have seen "Web 1.0" transition into "Web 2.0" with a "difference in scale, standardization, simplicity, and social incentives provided by web access turn a difference in degree to a difference in kind" (Allen, 2004). Whereas content on the WWW of the early 1990s was authored and disseminated by professionals and experts, contemporary content is driven by amateurs, by You and I in what Time magazine identified as person of the year in 2006 (Grossman, 2006),

> But look at 2006 through a different lens and you'll see another story, one that isn't about conflict or great men. It's a story about community and collaboration on a scale never seen before. It's about the cosmic compendium of knowledge Wikipedia and the million-channel people's network YouTube and the online metropolis MySpace. It's about the many wresting power from the few and helping one another for nothing and how that will not only change the world, but also change the way the world changes. And for seizing the reins of the global media, for founding and framing the new digital democracy, for working for nothing and beating the pros at their own game, *TIME*'s Person of the Year for 2006 is you.

1.5 An Encompassing Example: The GoNorth! Projects

Although this chapter is focused on the theoretical underpinnings of social computing and virtual communities, we believe that an example of the use of such theories in practice can help contextualize and solidify their importance.

Adventure learning (AL) is an approach to designing learning environments that provides students with opportunities to explore real-world issues through authentic learning experiences within collaborative learning environments (Doering, 2007; Doering and Veletsianos, 2008, in press). Grounded on theories of experiential and inquiry-based learning, the AL approach has been implemented in a social computing and virtual community context within the GoNorth! project (www.polarhusky.com). Specifically, in conjunction with the travels of Team GoNorth!, who annually dogsled throughout circumpolar Arctic regions, students from across the globe are faced with real-world problems and are involved in data analysis and interpretation, cross-cultural interaction, and collaboration with colleagues and experts. This endeavor brings together three million students every year and requires that the learner be situated at the heart of the project rather than at the periphery and be treated as a passive recipient of information. To this end, designing AL environments involves designers to evaluate the extent to which such environments consider issues related to interaction, constructivism, community, affordances, distributed cognition, CSCL, social presence, and the participatory web. These issues are explored individually below for the benefit of the reader, but it is important to note that analysis and (formative and summative) evaluation have been ongoing.

1.5.1 Interaction

Interaction is a fundamental process of social computing and, as such, is also an integral component of the GoNorth! project. Although the GoNorth! expedition team is on the Arctic trail, students are faced with numerous opportunities for social interaction. For example, they can read about the adventures of the team through the Weekly Trail reports posted on the online learning environment (see Figure 1.3), or view photographs and videos that document the adventures of the team. Students can also send personal notes to team members and engage in asynchronous interaction with both members of the team and other students worldwide. Finally, teachers, students, and classrooms can interact with each other, with the GoNorth! team, and with experts from across the world in synchronous and asynchronous discussion sessions.

1.5.2 Constructivism

The personal and social construction of knowledge has been an integral part of AL projects. Although avoiding extreme forms of unguided instruction (Kirschner et al., 2006), with regard to personal constructivism, AL allows learners to build on their prior understandings of being in the world. Constructivist understanding and knowledge is scaffolded in the GoNorth! project through three levels of inquiry and expertise: experience, explore, and expand. Each subsequent level requires increased student involvement while being increasingly less didactic. With regard

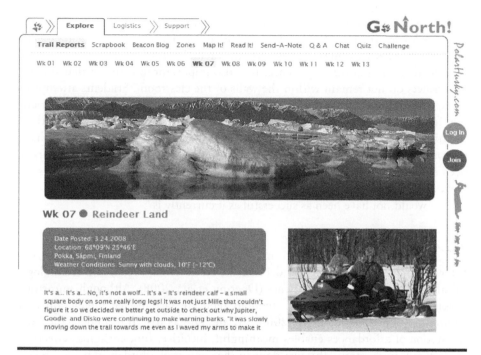

Figure 1.3 The GoNorth! team delivers a report from the trail.

to social constructivism, learners are frequently engaged in "expert chats" with participants from across the world. In these synchronous discussions, experts are invited to hold hourly discussions with learners. These discussions are moderated and classrooms are not only able to participate but also able to ask questions of the experts and contribute their own knowledge. The idea of "expertise" may need to be clarified here, as the GoNorth! project considers student participants to be experts as well. For example, students are given opportunities to contribute their artifacts and projects on collaboration zones that are then available to other students and teachers. The ability to develop and share artifacts not only gave students a sense of accomplishment but also instilled to them a sense of expertise and knowledge (cf. Doering and Veletsianos, 2008).

1.5.3 Community

The GoNorth! project has enabled the development of a community of educators, learners, environmentalists, and arctic explorers. These individuals, initially coming together for educational and environmental purposes, have formed a community of practice that goes beyond teaching and learning. Through mutual engagement with the GoNorth! project, participants, bonded by their concern of global environmental issues, have come to learn about each other on personal and intimate levels. The use of technology in the project has enabled participants not only to explore what it means to traverse the arctic by dogsled but also to actually get a firsthand feel of this experience. It is also important to mention that student experiences do not remain within the walls of the classroom: Students attempt to change their parents' behavior, parents request the continuation of the AL curriculum from the teachers, and teachers encourage each other to use the project in their teaching (Doering and Veletsianos, 2008). Overall, the GoNorth! expeditions have not only initiated the development of a community of learners but also actually been sustained by the community. Without the overwhelming participation, support, and encouragement of this (local and nonlocal) community, the GoNorth! project would not have been as successful as it currently is."

1.5.4 AL Affordances

Prior research has identified and defined the educational, social, and technological affordances of AL environments (Doering et al., 2008). Although numerous electronic learning environments tend to replicate face-to-face interaction and communication (Doering et al., 2008), AL design has strived to understand how the notion of affordances enables meaningful, intuitive, fun, engaging, effective, and efficient instruction. This was done in the context of affordances being broken down into three components (educational, social, and technological affordances), as described by Kirschner et al. (2004). Educational affordances refer to characteristics and features of a learning environment that enable and stimulate learners to meaningfully collaborate and engage with the instructional content; social affordances are characteristics of the environment that facilitate social interaction; and technological affordances are those features of the environment that enable the effective and efficient delivery of the learning experience. Although a full description of AL affordances is beyond the scope of this chapter, one example of harnessing technology for social interaction is provided below.

The online learning environment provides multiple options for learners to interact with and learn about each other. One such opportunity is provided by the Observations Map that provides a representation of the geographic origin of each interaction. By visualizing locations, learners are better able to situate themselves and their colleagues in the worldwide context within which their collaboration takes place.

1.5.5 CSCL

Collaborative learning lies at the heart of AL. Nevertheless, while collaboration occurs between geographically dispersed users, colocated students work and learn together using technological tools. For example, in our work with schools across the United States, we discovered that classroom participation in "Experts Chat" zones frequently takes place in the following manner (Doering and Veletsianos, 2008): Teachers gather their whole class together and use one computer connected to a projector as a shared display. The collaboration environment is projected on a large screen, and the class negotiates and contributes to the shared discourse. In this way, teachers and students collaborate face-to-face to contribute to a virtual collaborative environment. Although the reasons for using a shared display to collaborate on this task are varied, such reasons are part of CSCL study and include social, pedagogical, and organizational factors related to classroom participation in synchronous collaborative learning environments.

1.5.6 Social Presence

The use of real-time videoconferencing is limited in the GoNorth! project largely due to the simple fact that the expedition team is located in the Arctic and works under extreme conditions, using technology that may not always work (Miller et al., 2008). Nevertheless, social presence is an important component of the AL experience and, even in those instances where text is the predominant medium through which information is presented, the user experience is enhanced by media that attempt to bring to life what it means to live, breathe, and traverse the Arctic. For example, the learning environment makes heavy use of audio recordings, photographs, and short videos from the field. Such media enable students to get a feel for the strong and never-ending winds, to see the polar bear approaching the camp, to hear the howling dogs, and to get acquainted with the explorers and educators. Such was the degree of social presence attained in the 2004 expedition that online interactions translated seamlessly to real-life interactions:

> As we approach Baker Lake, Nunavut, we have not seen anyone else in 73 days. Across the horizon a jumping light can be seen as a snowmobile approaches us. I am on the front sled, so I stop the team and ski over to the individual dismounting his machine. I extend my arm and say, "My name is Aaron Doering. You have no idea how excited I am to meet you." The Inuit Elder from Igloolik, Nunavut, replies, "I know who you are; I recognize your voice from the Internet" (Doering, 2007, para. 1).

1.5.7 Web 2.0

Web 2.0 and AL share three important characteristics: user participation, "expert" vs. "novice" distinction, and rich user experiences. We have already discussed rich user experiences under different headings; therefore, this section will focus on user participation and "expert" vs. "novice" distinction—items that are inherently intertwined. The so-called Web 2.0 applications allow and enable users to contribute content and expertise: Individuals can comment on blog posts, edit wiki articles, aggregate and republish content, and bring together content from diverse applications to create new knowledge in the form of mash ups. AL participants can contribute artifacts to the learning environment, reuse AL content in their classroom, and even physically participate in AL expeditions. The idea of physical participation in such expeditions moves user participation to a new level, one that transcends the virtual world. Since 2005, four K-12 grade teachers (one per year) have participated in the AL expeditions carried out by team GoNorth!, bringing another level of interconnectedness among students, teachers, schools, and AL projects. Enabling teacher participation in such a project implies that teachers and intended users are treated as equals, as coparticipants in this endeavor to deliver environmental education, as experts in their own right, rather than as amateurs to whom content is delivered.

1.6 A Research Agenda

Although delineating the theoretical foundations of social computing and virtual environments is imperative, it is of utmost importance to also consider how to best engage in the research of social computing and virtual environments and what type of research may be most fruitful to expand our understanding of this field.

It is our view that research on social computing and virtual environments needs to be (1) empirical, (2) multidisciplinary, (3) multiparadigmatic, and (4) of a design-based nature. Empirical research is one that relies upon data (quantitative or qualitative) rather than opinion and claims. Multidisciplinary research refers to endeavors that function across academic disciplines and industries and encompass ideas from multiple and synergistic (but often diverse) fields. Multiparadigmatic research refers to those endeavors that do not necessarily subscribe to one single epistemology but rather employ multiple research methods (from different epistemologies) to understand phenomena. Finally, design-based research aims to understand activity in context and within authentic and naturalistic environments by developing "real-world" interventions and sharing knowledge among practitioners and researchers (Brown, 1992; Collins, 1992; The Design-Based Research Collective, 2003; Wang and Hannafin, 2005).

Given the above considerations, we believe that the following research questions will be of interest to the community and our colleagues:

- How do we initiate and foster *persistent* virtual communities?
- What frameworks can we develop to better understand social computing?
- How can social computing and virtual communities be used within sustainable development contexts?
- How is social computing and virtual community changing notions of citizenship, politics, and community organizing?

- How do technology affordances permit and constrain certain kinds of interactions?
- What does it mean to be a member of a community within which interaction is technology mediated?
- Where do informal learning tools fit in the formal education landscape?
- How does technology-mediated interaction shape structure, culture, and social presence?
- What combination of technologies, content, context, and instructional methods is appropriate for what kinds of instructional goals, teachers, and learners?
- How is emotion communicated online, and how do technological affordances shape emotional rules?
- What are the implications of social computing and virtual communities for individuals who have traditionally faced limited opportunities?

1.7 Conclusion

The purpose of this chapter has been to consider how three interrelated views in social computing—constructivism, situated and distributed cognition, and communities of practice—may be applied to the design, implementation, and evaluation of social computing applications. This chapter highlighted the value of collaboration, commitment, innovation, assessment, evaluation, communication, and interaction that underpins successful social computing applications. On the one hand, this work exemplifies the importance of strong theoretical foundations and highlights the value of our framework in contributing to a heightened understanding of social computing and virtual communities. On the other, this chapter serves as a call to researchers to investigate the possibilities and challenges concomitant to the rising popularity of virtual social networking technologies. Even though technology evangelists may praise the benefits of virtual socialization and technophobes might proclaim the end of face-to-face interactions, the fact is that social computing and virtual communities represent a complex phenomenon that has quickly come to the forefront of mainstream media and popular culture. Veletsianos and McCleary (2009), while discussing social networking sites as locales of youth socialization, emphasize the complexities of social computing by noting that although "virtual socialization provides an important site for socialization in informal social networks ... enabling youth to articulate and define their own sense of self and identity ... [virtual networking may be] another medium through which societal inequalities and the 'digital divide' are perpetuated."

Some of the most significant challenges to the study of social computing have to do with the difficulties related to the types of interaction that promotes user engagement, the lack of multidisciplinary research, and the focus on bipolar research paradigms. We hope that this chapter encourages practitioners and researchers to reach beyond their disciplines and collaborate in a quest to enhance our understanding of social computing and virtual communities.

References

Allen, C. (2004). Tracing the evolution of social software. Life with alacrity. Retrieved on October. 2008 from http://www.lifewithalacrity.com/2004/10/tracing_the_evo.html.

Bannon, L. J., and Schmidt, K. (1991). CSCW, four characters in search of a context. In J. M. Bowers and S. D. Benford (Eds.), *Studies in Computer Supported Cooperative Work* (pp. 3–16). Amsterdam: North-Holland.

Blumer, H. (1969). *Symbolic Interactionism: Perspective and Method.* Englewood Cliffs, NJ: Prentice Hall.

Brown, A. (1992). Design experiments: Theoretical and methodological challenges in creating complex interventions in classroom settings. *Journal of Learning Sciences, 2*(2), 141–178.

Brown, J. S., Collins, A., and Duguid, P. (1989). Situated cognition and the culture of learning. *Educational Researcher, 18*(1), 32–42.

Burbules, N. C. (2000). Does the Internet constitute a community? In N. C. Burbules and C. A. Torres (Eds.), *Globalization and Education: Critical Perspectives* (pp. 323–355). New York: Routledge.

Burbules, N. C., and Callister, T. A. (2000). Universities in transition: The promises and the challenge of new technologies. *Teachers College Record, 102*(2), 271–293. Retrieved on January 21, 2000 from http://www.tcrecord.org.

Cobb, P. (1994). Where is the mind? Constructivist and sociocultural perspectives on mathematical development. *Educational Researcher, 23*(7), 13–20.

Cole, M., and Engeström, Y. (1993). A cultural-historical approach to distributed cognition. In G. Salomon (Ed.), *Distributed Cognitions: Psychological and Educational Considerations* (pp. 1–46). New York: Cambridge University Press.

Collins, A. (1992). Towards a design science of education. In E. Scanlon and T. O'Shea (Eds.), *New Directions in Educational Technology* (pp. 15–22). Berlin: Springer.

Design-Based Research Collective, The. (2003). Design-based research: An emerging paradigm for educational inquiry. *Educational Researcher, 32*(1), 5–8.

Doering, A. (2007). Adventure learning: Situating learning in an authentic context. *Innovate-Journal of Online Education, 3*(6).

Doering, A., Lewis, C., Veletsianos, G., and Nichols-Besel, K. (2008). Preservice teachers' perceptions of instant messaging in two educational contexts. *Journal of Computing in Teacher Education, 25*(1), 45–52.

Doering, A., Miller, C., and Veletsianos, G. (2008). Adventure learning: educational, social, and technological affordances for collaborative hybrid distance education. *Quarterly Review of Distance Education, 9*(3), 249–266.

Doering, A., and Veletsianos, G. (2008). Hybrid online education: identifying integration models using adventure learning. *Journal of Research on Technology in Education, 41*(1), 101–119.

Doering, A., and Veletsianos, G. (in press). What lies beyond effectiveness and efficiency? Adventure learning design. *The Internet and Higher Education.*

Edigo, C. (1988). Video conferencing as a technology to support group work: A review of its failure. *Proceedings of CSCW* 88, 13–24.

Engeström, Y. (1987). *Learning by Expanding: An Activity-Theoretical Approach to Developmental Research.* Helsinki, Finland: Orienta-Konsultit.

Engeström, Y. (1999). Innovative learning in work teams: Analysing cycles of knowledge creation in practice. In Y. Engeström et al. (Eds.), *Perspectives on Activity Theory* (pp. 377–406). Cambridge, UK: Cambridge University Press.

Gibson, J. J. (1977). The theory of affordances. In R. E. Shaw and J. Bransford (Eds.), *Perceiving, Acting, and Knowing* (pp. 67–82). Hillsdale, NJ: Lawrence Erlbaum Associates.

Gibson, J. J. (1979). *The Ecological Approach to Visual Perception.* Boston: Houghton Mifflin.

Grossman, L. (2006). Time's Person of the Year: You. *Time Magazine*, December 13, 2006. Retrieved on October 12, 2008 from http://www.time.com/time/magazine/article/0,9171,1569514,00.html.

Grudin, J. (1994). CSCW: History and focus. *IEEE Computer, 27*(5), 19–27.

Gunawardena, C. N. (1995). Social presence theory and implications for interaction and collaborative learning in computer conferences. *International Journal of Educational Telecommunications, 1*(2/3), 147–166.

Gunawardena, C. N., and Zittle, F. J. (1997). Social presence as a predictor of satisfaction within a computer-mediated conferencing environment. *American Journal of Distance Education, 11*(3), 8–26.

Hoadley, C. M., and Pea, R. D. (2002). Finding the ties that bind: Tools in support of a knowledge-building community. In K. A. Renninger and W. Shumar (Eds.), *Building Virtual Communities: Learning and Change in Cyberspace* (pp. 321–354). London: Cambridge University Press.

Kim, A. (2000). *Community Building on the Web.* Berkeley, CA: Peachpit Press.

Kirschner, P., Strijbos, J., Kreijns, K., and Beers, P. J. (2004). Designing electronic collaborative learning environments. *Educational Technology Research and Development, 52*(3), 47–66.

Kirschner, P. A., Sweller, J., and Clark, R. E. (2006). Why minimal guidance during instruction does not work: an analysis of the failure of constructivist, discovery, problem-based, experiential, and inquiry-based teaching. Educational Psychologist, *41*(2), 75–86.

Lave, J., and Wenger, E. (1991). *Situated Learning: Legitimate Peripheral Participation.* Cambridge, MA: Cambridge University Press.

Miller, C., Veletsianos, G., and Doering, A. (2008). Curriculum at forty below: A phenomenological inquiry of an educator explorer's experiences with adventure learning in the Arctic. *Distance Education, 29*(3), 253–267.

Norman, D. (1988). *The Psychology of Everyday Things.* New York: Basic Books.

Olson, G. M., and Olson, J. S. (2008). Groupware and computer supported cooperative work. In J. J. Jacko and A. Sears (Eds.), *Handbook of Human-Computer Interaction* (pp. 545–554). Mahwah, NJ: Lawrence Erlbaum Associates.

Pea, R. D. (1993). Practices of distributed intelligence and designs for education. In G. Salomon (Ed.), *Distributed Cognitions: Psychological and Educational Considerations* (pp. 47–87). Cambridge, UK: Cambridge University Press.

Piaget, J. (1970). *Genetic Epistemology.* New York: Columbia University Press.

Rezabek, L. L., and Cochenour, J. J. (1998). Visual cues in computer-mediated communication: Supplementing text with emoticons. *Journal of Visual Literacy, 18*(2), 201–216.

Salomon, G. (1993). Editor's introduction. In G. Salomon (Ed.), *Distributed Cognitions* (pp. xi-xxi). New York: Cambridge University Press.

Schlanger, M. S., and Fusco, J. (2002). Teacher professional development, technology, and communities of practice: Are we putting the cart before the horse? *Information Society, 19*(3), 203–220.

Schwen, T. M., and Hara, N. (2003). Community of practice: A metaphor for online design? *The Information Society, 19*(3), 257–270.

Short, J., Williams, E., and Christie, B. (1976). *The Social Psychology of Telecommunications.* London: John Wiley and Sons.

Veletsianos, G., and McCleary, K. (2009). Social networking sites: socialization outside schools.

Von Glasersfeld, E. (1989). Cognition, construction of knowledge, and teaching. *Synthese, 80*, 121–140.

Vrasidas, C. (2000). Constructivism versus objectivism: Implications for interaction, course design, and evaluation in distance education. *International Journal of Educational Telecommunications, 6*(4), 339–362.

Vrasidas, C., and Glass, G. V., Eds. (2002). *Distance Education and Distributed Learning.* Greenwich, CT: Information Age Publishing.

Vrasidas, C., and Glass, C. V., Eds. (2004). *Online Professional Development for Teachers.* Greenwich, CT: Information Age Publishing.

Vrasidas, C., and Glass, C. V., Eds. (2005). *Preparing Teachers to Teach with Technology.* Greenwich, CT: Information Age Publishing.

Vrasidas, C., and Zembylas, M. (2003). The nature of technology-mediated interaction in globalized distance education. *International Journal of Training and Development, 7*(4), 1–16.

Vrasidas, C., and Zembylas, M. (2004). Online professional development: Lessons from the field. *Education and Training, 46*(6/7), 326–334.

Vygotsky, L. S. (1978). *Mind in society.* Cambridge, MA: Harvard University Press.

Wang, F., and Hannafin, M. J. (2005). Design-based research and technology-enhanced learning environments. *Educational Technology Research and Development, 53*(4), 5–23.

Wenger, E. (1998). *Communities of Practice. Learning, Meaning and Identity.* Cambridge: Cambridge University Press.

Wenger, E., McDermott, R., and Snyder, W. M. (2002). *Cultivating Communities of Practice: A Guide to Managing Knowledge.* Cambridge, MA: Harvard Business School Press.

Zembylas, M., Vrasidas, C., and McIsaac, M. S. (2002). Of nomads, polyglots, and global villagers: Globalization, information technologies, and critical education online. In C. Vrasidas and G. V. Glass (Eds.), *Distance Education and Distributed Learning* (pp. 201–224). Greenwich, CT: Information Age Publishing.

Chapter 2

An Overview of Social Networks

Gindo Tampubolon

Contents

2.1 If There Was Ever a Beginning, It Must Be Mom and Dad...

The focus on relationships in our attempt to make sense of the world around us is probably innate to all of us. I will relate a story. A three-year-old immigrant girl enrolled in a nursery school in the Northwest of England was speaking in her native language at home with her parents; she certainly did not called her father "Dad" nor her mother "Mom." After a couple of weeks in school, one day she came home and did her usual things, which was to play with her little brother and her parents. In the course of playing, matter-of-factly and to their surprise, she called her father "Dad" and her mother "Mom." The parents, in relating the story, reasoned that she probably observed parents of her schoolmates at the school gate and correctly observed that these parents' relationships with her mates were no different than her own parents' with her. So if these parents were called Dads and Moms, surely what she had at home were also Dad and Mom. To social network analysts, the little girl has just deduced regular equivalence (relating two roles, i.e., parents and children) from observing repeated and multiple relationships (Wasserman and Faust 1994, Chapter 12).

The place of family in this genesis story of social network analysis is not far-fetched. Freeman (2004) traces the ideas of social network analysis to the structural sociology of Auguste Comte (2004, p. 11–14). He concludes Comte's contribution, quoting his emphasis on the fundamental role in a family of the couple, or "dyad" as it is called in modern social network analysis. The structural sociology of today focuses on relations among concrete agents or actors such as individuals or organizations. It contrasts with those that focus on ideas or conceptual maps or on statistical models of variables gathered through social surveys. A focus on dyads or ties that connects two social actors, as opposed to a focus on individual actors, is one among four fundamental features of social network analysis as it is known today. To recap from Freeman (2004):

1. Social network analysis is motivated by a structural intuition based on ties linking social actors.
2. It is grounded in systematic empirical data.
3. It draws heavily on visual imagery.
4. It relies on the use of mathematical and/or computational models.

This chapter aims to provide a brief overview of empirical social network analysis by elaborating on the four common characteristics of existing analyses. It also points to an emerging science where social network analysis plays a fundamental role in bringing together a deeper understanding and connections between the increasingly merged offline and online social processes.

Although issues in social network analysis can be traced back to enduring relational issues about family, friendship, neighborhood, and society, these issues take on a different character today with the ubiquity and pervasiveness of the online media, especially the World Wide Web and mobile communication. The different character only serves to indicate that social networks and social network analysis are at a ferment phase where new issues are being redefined and new conceptions are being forged. A definitive overview of the field is thus less useful than a broad brush and selective overview. This choice allows discussions of the wide variety of fields that have borrowed the metaphor of social networks, to apply ideas and tools from social network analysis, and to fundamentally contribute to the subjects of social network analysis (i.e., what can reasonably be seen as a network and the new tools and methods for understanding processes operating in such networks.) Freeman's identified features serve as a tentative organizing device that will be used when helpful. We will also find more emphases on recent developments—not, one might hope, as a victim of fads and fashions but out of enduring concern that social network analysis is fundamentally about "social" relations. Recent developments have begun to raise new questions about sociability, as I hope to show. What fields do I discuss and what fields are necessarily left out? In a short, selective, and ultimately subjective review like this, certain fields are easily justified as relevant and interesting. Such is the case of friendship and its influence on the behavior of a group of friends. Other fields deserve special and separate treatment due to their specific vocabulary and emphases on certain substructures or subnetworks. Such is perhaps the case with chemical and biological networks. Again, this is all evidence to support the idea that social network analysis has found an even wider application as we increasingly adopt the metaphor of networks and the methods of social network analysis in understanding social relations and processes around us.

2.2 Freeman's First Characteristic: What Is Social about Social Network Analysis?

It is a remarkable feature of social network analysis that once the metaphor of network is grasped, many social processes or behaviors of interest take on a more intricate pattern. It is as if relations and objects that were previously out of focus slowly come into sharper view. A network—or, mathematically, a graph—is defined simply by the set of vertices and the set of edges between these vertices. Vertices can be persons, such as mom, dad, teachers, schoolmates, or friends; and edges can signify caring, visiting, or sharing a ride home. Figure 2.1 presents a network of friendships among workers (Ws), supervisors (Ss), and inspectors (Is) in a bank wiring room—the subject of a famous study that gives us the "Hawthorne effect" (Roethlisberger and Dickson, 1939). The actors are workers, supervisors, and inspectors, and the edges capture friendship ties among the actors.

Social network analysis is motivated by the intuition that social processes, such as sharing a ride home or friendship, are better understood as contextual or as network processes. For instance, a friend of a friend often becomes a friend. In social network analysis, this is known as transitive structure or triangle network motif. Figure 2.2 gives three examples of triads commonly found in empirical networks: outgoing triad, attracting triad, and complete (friend-of-a-friend-is-a-friend) triad.

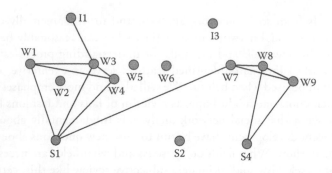

Figure 2.1 Network of friendships among workers (ws), supervisors (ss), and inspectors (is) in a bank wiring room, the subject of a famous study that gives us the "Hawthorne effect" (From Roethlisberger and Dickson, *Management and the Worker,* 1939).

An example may serve to illustrate how fruitful social network is as one perspective in understanding social processes or behavior. The precursor to modern social network analysis, which can be traced back to Jacob Moreno's sociometry of the 1930s, concerns itself with "obtaining, by application of quantitative methods, [an understanding] about the evolution and organization of groups and the position of individuals within them (Freeman 2004: p. 37)." For instance, "an epidemic of runaways among the girls at the Hudson school could be explained by chains of social ties that linked all those who had left" (in Freeman 2004: 37). Borgatti et al. (2009) recount this classic study, "The reason for the spate of runaways had less to do with individual factors … [personalities and motivations] than with the positions of the runaways in an underlying social network."

Figure 2.3 shows nine out of fourteen girls who ran away and their feelings of attraction or mutual attraction. They lived in two cottages on the left and the right of the network. The runaways were labeled with light blue colors and those who stayed were not labeled. All the girls ran away over a period of only two weeks, a rate that was unprecedented. For Moreno, these links provided "channels for the flow of influence and ideas." The links were, in this way, more influential than their own individual personalities and motivations.

Figure 2.2 Three examples of triads commonly found in empirical networks: outgoing triad, attracting triad, and complete (friend-of-a-friend-is-a-friend) triad.

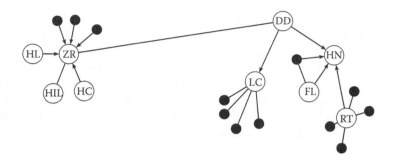

Figure 2.3 Runaway girls (white nodes) in upstate New York in 1932 and their feelings of attraction or mutual attraction.

The social ties as scaffolding for individuals' decisions have modern-day and everyday resonance. A piece of news on the British Broadcasting Corroboration Web site dated January 23, 2008, reports on the suicides of seven teenagers in the area of Bridgend, South Wales. One explanation that concerned people try to understand and discount is the role of social networking Web sites. Friends, parents, teachers, the police, the local member of parliament, local councilors, and local health trust members are anxious to find out whether the teenagers had made contact with each other through e-mail or Web links. One old as well as novel facet of this story relates to social network analysis. It surely needs examination; that old aspect of social ties in the teenagers' decisions is one possible explanation. This is similar to the story of the girls from the Hudson school some decades ago. A novel aspect to this is the role of the World Wide Web as a medium where social ties are struck, built, and maintained. Perhaps the World Wide Web pervades all four fundamental features of social network analysis as identified by Freeman.

Some of the research questions in social network analysis emphasize this enduring interest in social processes:

- Do friends of a friend tend to become friends as well? Do scientific collaborations extend in such a way as to facilitate direct collaborations between previously only indirectly connected scientists? Variants of this transitivity process abound in different settings.
- How much does background similarity matter in friendship? Does ethnicity matter? Does gender matter? Various aspects of homophily or similarity may be at work that can lead to an observed preponderance of links among those with similar background characteristics.
- What regularities at the societal level, say a tendency for marital educational homogamy, can be explained by social interaction at the local and individual level? Can regularities in consumption activities, say the rise of the iPod as a must-have item for self-respecting urban teenagers, be explained in terms of social networks of teenagers?

In order to begin answering these kinds of questions from a relational perspective, social network analysis turns to a specific kind of empirical data, that is, network data. Network data tend to look like a collection of pairs of actors such as (worker A, supervisor B.) In this way, the data for Figure 2.1 would look like the following: (I1, W3), (W1, W3), (W1, W4), (W3, W4), (W1, S1), (S1, W3), (S1, W4), (S1, W7), (W7, W8), (W7, W9), (W8, W9), (S4, W8), and (S4, W9). Although these ties can also be represented by a matrix of zeroes and ones representing present or absent friendship, for a large number of actors, pairs like the above are more efficient.

2.3 Freeman's Second Characteristic: Describing Whole Network and Ego Network

Whole network and ego network are two designs of collecting network data. In whole network design, the purpose is to collect actors and relations (or vertices and edges) within an a priori specified network boundary. This is often facilitated with an available roster of vertices. The boundary is, for instance, a school or classroom. The boundary that limits the network vertices can be chosen on either theoretical or pragmatic grounds. Until very recently, the case of network analysts and network data is like the president of the United States and world events: one is at the mercy of the other.

Although vertices of the network are bounded pragmatically to one setting, the edges or relations can go beyond that setting. To continue with our example with a school-bounded network, one may study friendship on the school ground *and* relationships outside school, for example, going to the movies together. In fact, this is one kind of interesting investigation with *multiple* networks where the vertices set is the same and the edges set is multiple. The substantive questions raised here, for instance, is whether affections or relations spills over from one domain to another. To give a more contemporary take on this, one can ask whether networks in this life spill over onto networks in *Second Life*™ (http://secondlife.com) or vice versa.

The problem of setting network boundaries should not be underestimated, as in practice this problem is inextricably linked with what could substantively be inferred, especially in the snowball sampling method of data collection. In this method, the decision to pursue the depth of sampling substantively limits the possible network boundary. However, sampling methods such as snowball sampling (as opposed to those based on complete roster of actors or whole network design), which can be seen as a restricted method to uncover a whole network, can better capture the process of influence underlying the formation of the said network.

We now have access to technology-driven raw relational data, which is not entirely unrelated to how technology has been changing relations in society. Mobile phone and SMS (short messaging service) logs, personal Web sites, and, especially, weblog links and Wikipedia links just to name a few are churning up a wealth of network data. There was a brief period where these threatened to overwhelm social network analysts, as each analyst writes a unique Web crawler to harvest the data. The situation has improved tremendously with the recent Web services–oriented applications around where the application programming interfaces are released. For instance, the European Patent Office recently published the Open Patent Services Web service interface at http://ops.espacenet.com. It facilitates the automatic collection and analysis of networks of technology inventors in many fields of technology across different countries because the European Patent Office grants patent protection in many countries.

Even with this wealth of possibilities, we are, of course, nowhere nearer in understanding what they all mean in terms of sociability, identity, authority of news, and nature and codification of knowledge. One thing is certain on the face of this technology-driven wealth of possibilities: We will stay where we are in our understanding if we fail to deploy methods and tools of network analysis and do not draw from what we know about social network.

In contrast to whole network data, ego network data provide only partial views of networks based on each actor's local or direct ties. Before we are all attracted to this wealth of data and abandon ego network data, we should say that it endures in important studies. Ego network data, collected often through standard sample surveys, furnish an analyst with data about an individual and his or her friends, acquaintances, associates, and contacts. With these, facets of the structure of sociability—such as marital endogamy and its role as empowering certain behaviors or constraining certain others—are often the subject of investigation. Nationally representative samples of ego networks are regularly collected in industrial countries. These provide snapshots of sociability and its changes over time. As such, this serves as a useful background against which to compare the online network.

Although whole network and ego network are often taken as two distinct designs, in practice or in analysis they are not mutually exclusive. One can take a whole network and go on to derive ego network indices and proceed with the analysis. Conversely, albeit in very specifically dense setting, one can sample enough ego networks and go on to derive a whole network (though often patchy) by inferring the contacts as sample egos.

Ego network data can be satisfactorily analyzed and the results communicated with standard methods and materials of quantitative science. However, whole network data require novel methods and materials to convey the results of social network analysis. These methods and materials are predominantly visual.

2.4 Freeman's Third Characteristic: Visual Imagery

The role of visual imagery in social network analysis cannot be overestimated. Visuals allow standard notions of social network analysis and key concepts in the substantive areas applying it to be shown succinctly and directly. Notions such as centralization, connectedness, core, or periphery are not only methodological notions but also translatable to substantive concepts, including for instance power, popularity, influence, elitism, or disadvantages. Visual imagery allows a nearly straightforward translation of "everything is connected to everything else" onto paper or screen or, increasingly, animation. Again, social network analysis has benefitted from the technological and methodological innovations, for instance, in the visualization and explication of change. Networks can be examined in the range of software, from a standalone software for network analysis like Pajek (Batagelj and Mrvar, 1999) through a library in an open source statistical computing environment like sna in R (Butts, 2009; R Development Core Team, 2008) to a Java class framework like JUNG (http://jung.sourceforge.net, accessed February 23, 2009.) Phase transition or critical mass or critical threshold, which separates two qualitatively different regimes, can easily be made visual.

In the other direction, substantive concerns of social network analysis and issues of cognition have raised challenges and drive further methodological developments in the visualization of networks. Novel methods and algorithms for visualizing hierarchy and power add to the synergy between social network analysis and the technology of visualization. This challenge was intensified recently with the wealth of network data arising from technology-driven records of a whole network of relations. Not only visualizing essentially static concepts such as popularity, visualizing dynamic notions such as continuous contagion or influence is an added challenge. The difficulty in visualizing dynamics goes with the complexity in modeling dynamics. For a recent example in modeling dynamics, Spinellis and Louridas (2008) analyze the evolution of entries in the online encyclopedia Wikipedia. They have kindly posted the raw data and C++ code for their updated analysis at http://www.dmst.aueb.gr/dds/sw/wikipedia (accessed February 23, 2009).

2.5 Freeman's Fourth Characteristic: Methods for Network Dynamics and Behavioral Change

If the promise of social network analysis is the understanding of behavior in its relational context, then modeling behavioral change simultaneously with network dynamics is the holy grail of social network analysis. There has been a fresh round of works on these twin issues in the past few years. With the help of real settings, the two main questions of influence and selection can be discerned here.

Social influence is thought to operate when friendship network facilitates one member to change the behavior of a friend by virtue of links to the network. Say two friends are part of a network: One is a smoker, the other is not. After some time, the other friend becomes a smoker too as they remain within the same network of friends.

Social selection is distinct from social influence: Instead of change in behavior as the consequence of being in a network (cause), the process is reversed. A smoker who is not part of a network seeks out a friend (and friends of the friend in the network) who also smokes. Smoking here is the driving force for selecting the friendship network to join.

Rigorously teasing out the two processes of influence (that change behavior) and of selection (that change network structure) can quickly become an involved exercise in sophisticated statistical modeling.

This teasing out is not an idle exercise, as illustrated by two widespread and important applications in the diverse fields of public health and marketing. Even if only implicit or only based on limited evidence, public health interventions among teenagers rely on the efficacy of social influence delivered through networks of friends. Smoking awareness intervention relies on the crucial role of friends in influencing what choices of healthy lifestyles are available to teenagers and what choices are simultaneously deemed socially desirable and generally healthy. Sexually transmitted disease interventions also rest on the idea that social influence through networks of friends is efficacious.

The epitome of social influence in consumer marketing fields is perhaps captured in the whole practice of viral marketing. There are many stories of marketing executives depicting networks of target consumers in order to identify centers, hubs, or gateways. Such positions are believed to be positions of influence that can change the behavior of those who are linked to them in the network. In the world of limited resources and short durations, positions with potentials such as these are very much sought after.

Examples of social selection are perhaps more readily available though not as memorable. After all, hobbies and other activities that we follow with some intensity are often done with a group of "like-minded" individuals. Begging the readers' pardon for the pun, knitting, though often seen as a solitary activity, readily induces its practitioners to find others who enjoy the same activity and thus knit a social network. This social selection is clearly not always amateurish or voluntary. Any institutionalized profession, such as the various branches of the sciences, organizes meetings, conferences, and workshops to facilitate, among other things, this selection into its professional network.

On further reflection, however, networks and behaviors of their members are both cause and effect. This, of course, is not much help to public health officials, marketers, and volunteer officers of a profession. In the very important case of public health, however, recent efforts at modeling and teasing out the direction of causation is a welcome development.

Much of the heat and light that attract many scientists and analysts into social network analysis recently perhaps has to do with the rediscovery and extension by physicists on the phenomenon of small world and its underlying mechanisms (Durrett, 2007). To explain the small world phenomenon, it is perhaps useful to use a counterintuitive starting point such as an extremely large network. In giant networks such as the world's population or the World Wide Web, any two random individuals or two random sites can be connected with only a small number of intermediate contacts; the number is much lower than a naive guess might suggest.

The World Wide Web, in this sense, displays a small world property. In addition, the mechanism giving rise to the observation has been explained primarily in terms of local contacts plus a small fraction of random long-range contacts. Herein lies the attractiveness of the mechanism as it makes the connection between microlevel interaction and macrolevel regularity. The small world phenomenon has been observed in a wide variety of settings, such as movie actors' collaborations, mathematicians' collaborations, and the World Wide Web as a graph, to name a few. The small number of connections reported in large populations was observed systematically in an experiment conducted by psychologists Jeffrey Travers and Stanley Milgram; its academic report appeared for the first time in 1969 in the journal *Sociometry*.

Often mentioned in the same breath with the small world network is the mechanism of power law neighbors distributions and preferential attachment. It has been observed that the World Wide Web (with its hypertext protocol) has a probability degree distribution that obeys power laws: $p_n \sim Cn^{-g}$. In this distribution, the probability of observing an actor with n neighbors is proportional to its power. It is often observed that g ranges from 2 to 4 in the World Wide Web, the physical router network that makes up the Internet, the movie actor network, the collaboration network of authors, and the sexual networks of nearly five thousand Swedes (Durrett, 2007). A mechanism that has been proposed to explain this large-scale regularity is described in terms of social processes of preferential attachment. As a new vertex appears in the form of new Web sites or routers, it is likely to make connections with an already popular router or Web site. Thus, the popular ones get ever more popular because they are more preferred when a new connection is made. This preferential attachment is of course a well-known observation in the sociology of science, as discussed in 1968 by Robert K. Merton.

Although large-scale regularities observed on the World Wide Web or online connections can sometimes be traced back to old ideas that are well known and well documented in the social sciences, large-scale networks whether social or technological of course raise novel questions. The trick is to build on what is known to peer into what is yet unknown. This is clearly the point of departure for a new Web science as advocated by Tim Berners-Lee and his associates.

2.6 Toward a Web Science

The perspective that the World Wide Web is a mathematical graph has given us a deep understanding of contemporary technology networks and large-scale social networks. However, a purely mathematical and technological perspective pushes a large part of the vista into our blind spot. The proponents mention two specific instances that require the adoption of multiple perspectives, including social perspectives. Increasingly, our Web crawler or Web service clients are following links and thus constructing networks that are influenced by our own dynamic interactions with the sites (as, for instance, recorded with the help of cookies) encountered along the way. Our interactions with the sites are neither solely nor primarily technological; hence the online graph we construct for social network analysis is considerably influenced by factors other than technology. Little is known about this dynamic large-scale graph created "on the fly" in increasingly more cases.

A second instance is expounded using the relation between MediaWiki software, on the one hand, and, on the other, Wikipedia (or one of the more famous World Wide Web encyclopedias.) The point of the relation between the two is apparent only by virtue of the fact that the same software was used to build other applications that simply failed; hence, we do not know much about them. The success of Wikipedia is, then, only assured because of the right combination of the technology, the nature of the service, i.e., knowledge codification by volunteers in a social process (studied, for instance, by people who study social movements), and perhaps something else.

If new sociability and social networks and their encouragement by the World Wide Web and other online media are to be understood, then a new science is needed, that of Web science. This new science focuses not only on the protocols, services, and applications but on their social processes as well. If friendships, acquaintances, associates, contacts, and other social relationships are to be nurtured rather than neglected to atrophy in the ubiquitous World Wide Web, then Web science is needed to understand them in their changing social and technological contexts.

Some of the new research questions in social network analysis brought about by the World Wide Web and online media include

- What is the contour of the new *sociability* mediated by the World Wide Web and online media such as Flickr™, Friendster™, MySpace™, Facebook™, and Twitter™? If friendships on the World Wide Web can amount to hundreds and thousands (as opposed to tens and twenties of offline friends), is there an emerging form of sociability that is qualitatively different from offline friendships that involve recurrent face-to-face meetings?
- With a historical precedent of extensive transportation infrastructure stretching friendships to overcome local geography and extending friendships to the other ends of town (Fischer, 1982), with the World Wide Web the spectrum of sociability is being stretched to overcome geography even more and to transcend time. How should the spectrum of sociability be extended when the categories of friendship, acquaintances, and associates are evidently insufficient?
- More and more resources on the World Wide Web are being tagged or classified by users leading to what is known as "folksonomies" or folksy taxonomies. The former is obviously not in the Oxford English Dictionary yet though it clearly occupies a certain position in the World Wide Web. The creation of these tags and the extraction of meaning from these tags are not removed from the specific social context or social networks, online and offline, where they originated and circulated. How should we understand these sense-making online communities?

2.7 Conclusion

The World Wide Web and other online media are renewing and extending the interests in social network analysis. I have attempted a prospective overview with emphases on making direct connections with past studies, making connections between online and offline social networks, and, more importantly, pointing to one, among many, potentially exciting ways of looking scientifically at social networks in the ubiquitous World Wide Web.

More and more social networks' data will be coming online from the past, as archives are digitized and as transaction data in suitably anonymous forms are brought onstream. It is clearly a boon for social network analysis. Other groups of professionals, such as the computer scientists and researchers working on the intersection of human-computer communications, also welcome this development as more and more of the analysis and communication will involve intense interactions among users, analysts, developers, and computer systems.

There is little doubt that today is also an exciting period for mathematicians, physicists, and computer scientists who are working on understanding the structural and dynamic properties of especially large networks. In the words of one of the physicists (Mark Newman) to one of the mathematicians (Rick Durrett, Random Graph Dynamics p. 2), "I think there's room in the world for people who have good ideas but don't have the rigor to pursue them properly [MN]—makes more for mathematicians to do." There seems to be an open field where researchers from different fields in science and the humanities can collaborate to explore and understand the increasingly dense, ubiquitous, offline, and online networks embedding us.

2.8 Further Readings

In a subjective and prospective overview like this, there are bound to be omissions. This list is definitely idiosyncratic, but I will attempt to say a few words for each selection.

Adams and Allan, 1998. This collection of works on the social context of friendships in industrial societies is slightly dated though it contains some discussions on recurring issues, such as friendships and technological change. See also Spencer and Pahl (2006).

Buskens, 1999. This thesis employs one variant of social network analysis where another method is intensely used and other substantive concerns are the focus of investigation. The social network provides a structure in which game theory is used to analyze the formation and maintenance of trust.

Caldarelli and Vespignani, 2007. This is another one from the physicists with a much more circumscribed perspective of complexity. Thus, it has much similarity to that of Dorogovtsev and Mendes in its focus on self-organized dynamics. It also discusses applications in natural science and finance. The latter should perhaps not be seen as an unequivocal success given the mess the global financial system is in at the moment.

Carrington, Scott, and Wasserman, 2005. This claims to complement Wasserman and Faust (1994), and it does so admirably. In addition to picking up where Wasserman and Faust left off, it quickly gains speed to cover the state of where things were in early 2000.

Dorogovtsev and Mendes, 2003. This is one of the many recent books by physicists on the formal characteristics of primarily large-scale networks, though some characteristics discussed are also observed on the smaller scale. It covers largely the same ground as Durrett (2007) but also includes topics closer to the interests of physicists, such as percolation, the Ising model, and self-organized criticality in complex systems.

Durrett, 2007. A probabilist's take on the field occupied recently by physicists working on social networks. There are many refinements, corrections, and proofs of simulation-based results that physicists have brought to public, if not academic, understanding. They are all then put on a more secure mathematical footing. These include six degrees, small world phenomenon, power laws, preferential attachments, Erdös-Rényi random graphs, average path length of giant components, and phase transitions under attacks, culminating in the characterization and proof of the dynamic random graph of Callaway, Hopcroft, Kleinberg, Newman, and Strogatz.

Fischer, 1982. This is still one of the important resources on urban sociology and contemporary friendships for both its methodological rigor and interesting substantive insights. Friendships and different forms of sociability in urban settings and pervasive technologies are the subject of this important book.

Freeman, 2004. A history of the field by one of its early and accomplished scholars, this work thus has a nearly autobiographical character to it. This does not detract at all from its value. It is subtitled "A Study in the Sociology of Science," a subject studied by Robert K. Merton that gave us preferential attachment.

Friedkin, 1998. The author provides a thorough exposition of one alternative of understanding social influence within social networks. Borrowing ideas from social psychology, influence and the formation of consensus are the two social processes that are explicated in this work. There is insufficient detail on the other side of the coin, i.e., social selection, perhaps because this is taken to be largely understood relative to the subject of social influence.

Kanehisa, 2000. This biologist's take on post–human genome biology relies greatly on network analysis of large-scale biological and chemical data that are increasingly available. Some of these data are noted and described—a fascinating area of application.

Spencer and Pahl, 2006. The authors present a thoroughly qualitative study of contemporary friendship in Britain complemented with a limited quantitative and descriptive analysis.

Wasserman and Faust, 1994. This is the canonical book for social network analysis. It is a compendium of all major methods used in the field until the mid-1990s. For example, it has an extensive and formal treatment of triads, an example of which is "friend of a friend is a friend." Today, triads are resurrected as network motifs.

Periodicals. The standard periodicals include *Social Networks,* but increasingly *Physical Review E*, a few of the *IEEE Transactions*, and many of the professional magazines such as *Communications of the ACM*.

References

Adams, R.G. and Allan, G. 1998. *Placing Friendship in Context.* Cambridge University Press. Cambridge.

Batagelj, V. and Mrvar, A. 1999. "Pajek – Program for large network analysis." *Connections.* Vol. 21, No. 2: 47–57.

Borgatti, S.P., Mehra, A., Brass, D.J. and Labianca, G. 2009. "Network analysis in the social sciences." *Science*. Vol. 323: 892–895.

Buskens, V. 1999. *Social Networks and Trust.* Interuniversity Center for Social Science Theory and Methodology. Utrecht, The Netherlands.

Butts, C.T. 2009. "The sna package." http://erzuli.ss.uci.edu/R.stuff/. Accessed 23 February 2009.

Caldarelli, G. and Vespignani, A. Editors. 2007. *Large Scale Structure and Dynamics of Complex Networks: From Information Technology to Finance and Natural Science.* World Scientific. Singapore.

Carrington, P.J., Scott, J. and Wasserman, S. 2005. *Models and Methods in Social Network Analysis.* Cambridge University Press. Cambridge.

Dorogovtsev, S.N. and Mendes, J.F.F. 2003. *Evolution of Networks: From Biological Nets to the Internet and WWW.* Oxford University Press. Oxford.

Durrett, R. 2007. *Random Graph Dynamics.* Cambridge University Press. Cambridge, UK.

Fischer, C.S. 1982. *To Dwell among Friends: Personal Networks in Town and City.* The University of Chicago Press. Chicago, IL.

Freeman, L.C. 2004. *The Development of Social Network Analysis: A Study in the Sociology of Science.* Empirical Press. Vancouver, BC, Canada.

Friedkin, N. 1998. *A Structural Theory of Social Influence.* Cambridge University Press. Cambridge.

Kanehisa, M. 2000. *Post-Genome Informatics.* Oxford University Press. Oxford.

Moreno, J. 1934. *Who Shall Survive?* Nervous and Mental Disease Publishing Company. Washington, DC.

R Development Core Team. 2008. "R: A language and environment for statistical computing." R Foundation for Statistical Computing. Vienna, Austria. http://www.R-project. org. Accessed February 23, 2009.

Roethlisberger, F.J. and Dickson, W.J. 1939. *Management and the Worker*. Harvard University Press. Cambridge, MA.

Spencer, L. and Pahl, R. 2006. *Rethinking Friendship: Hidden Solidarities Today*. Princeton University Press. Princeton, NJ.

Spinellis, D. and Louridas, P. 2008. "The collaborative organization of knowledge." *Communications of the ACM*, Vol. 51, No. 8: 68–73.

Wasserman, S. and Faust, K. 1994. *Social Network Analysis: Methods and Applications*. Cambridge University Press. Cambridge, UK.

Wikipedia growth dds. http://www.aueb.gr/dds/sw/wikipedia. Accessed February, 23, 2009.

Chapter 3

Research Methods 2.0
Doing Research Using
Virtual Communities

Paul Cairns and Mark Blythe

Contents

3.1 Virtual Communities as a Resource for Research

Virtual communities offer an unprecedented access to new forms of primary data in many fields of inquiry. In particular, user-generated content on sites such as YouTube™, Facebook™, and MySpace™ provides primary data about both topics of interest to users of those sites as well as the responses of the online community to the posted content. For example, shortly after the launch of the iPhone™ 3G, there were literally thousands of video clips uploaded to YouTube that offered demos and reviews of the new product (Blythe and Cairns, 2009). Additionally, many people posted comments and ratings about these videos. Such materials could provide a rich resource both in researching the community of people who buy or use the iPhone 3G and in the design of future products.

However, both the quality and the quantity of the available data offer substantial obstacles to existing research methods and therefore bring into question the value of research based on this data. Additionally, the very dynamic nature of user-generated content sites makes it impossible to be exhaustive and presents something of a moving target to the researcher.

In this chapter, we consider how existing quantitative and qualitative methods may be applied to researching the primary data offered by virtual communities. Specifically, we focus on analyzing YouTube content, and we discuss how different methods apply, what their limitations are in this context, and how those limitations may be overcome through adaptations of the methods or even new methods.

The existing methods considered are statistical analysis, content analysis, and grounded theory. These methods are used to study not just the virtual community embodied in YouTube but also people's relationships with technology as reflected in the videos and their associated comments. It becomes clear that each method has something to say about understanding YouTube content and could say more if developed for this sort of data.

What is also clear, though, is that in some sense these methods miss some of the key sociological phenomena around the videos. In previous work (Blythe and Cairns, 2009), we found that the most popular video related to the iPhone 3G was the "Will It Blend?" video where an iPhone 3G is blended to dust in a parody drawing on 1960s game shows and scientific documentaries, among other things. The sheer numbers of people viewing this video and commenting on it, more than 2.5 million and 10,000, respectively, show that this video has an enormous impact; but being a single video, it is hard to use existing research methods to account for this impact.

Critical theory is being increasingly used in human-computer interaction (e.g., Bell et al., 2003; Bertelsen Pold, 2004; Sengers et al., 2005; De Souza, 2005; Blythe et al., 2008; Satchell, 2008) and offers researchers the opportunity to analyze cultural artifacts to provide insight into the social and cultural phenomena. In this chapter, we offer an introduction to critical theory and suggest a method for applying it to the study of user-generated content.

In order to demonstrate the methods discussed in this chapter, we have used a videogame, *The Shadow of the Colossus* (Sony, 2006), as the focus of our research. This is in part because videogames are a significant cultural phenomenon and area of substantial research interest (e.g., Juul, 2005) and in part because this particular game, released in 2005/2006, is relatively old and so presents something of a more stable topic for study and for applying the different research methods. We are therefore not considering in this chapter the particular challenges presented by rapidly emerging and evolving topics such as those related to game or product launches or as a consequence of news stories. (For example, at the time of writing, it is less than twenty-four hours since a plane crash-landed in New York's Hudson River without loss of life, and there are several hundred uploaded videos on YouTube and no doubt will be many more before this chapter is finished.)

3.2 *Shadow of the Colossus*

Shadow of the Colossus was released in Asia and the United States in 2005 and elsewhere in 2006 (Wikipedia, 2009b). The player controls the character Wander and must slay sixteen colossi, enormous creatures that inhabit a beautiful, yet deserted, land populated in addition to the colossi by only a few birds and animals. There is also an implicit backstory to the events of the game, which unfolds during the course of the game, and a substantial end sequence. This makes the game different from many others in that there is no persistent action in the game: Much of the activity of the player is riding Wander's horse, Agro, through this empty landscape to find the next colossus. The game is intended to (and does) feel like sixteen "boss" levels of a more traditional game but without ordinary levels in between. Indeed because a colossus presents a seemingly impossible challenge to Wander, the contrast between the intensity of fighting and the relaxed, occasionally aimless searching for the colossus is all the more marked.

The production values of the game are high: There is an excellent rendering of the landscape and the colossi; a powerful third-person perspective that, through camera shakes and blurs, adds to the impression of size and power of the colossi; and an unobtrusive, yet atmospheric soundtrack that only plays during encounters with a colossus.

The game has sold well, having had a wildly popular viral marketing campaign (Pspimp, 2009) and building on the cult success of *Ico*, a previous release by the same developers. It also has received many awards, including for its artwork, music, and overall design (Wikipedia, 2009b).

The search was done in YouTube by using the search term "Shadow of the Colossus." It is immediately clear that an important subset of the videos returned are clips showing "speed runs" where a player has managed to kill a colossus in impressively quick times. These use the timed attack feature that is unlocked when players complete the game. To look at the relationship between a subsample of video clips constituting part of a larger sample, the secondary search term "Shadow of the Colossus timed attack" was used. Also, to see changes over time, each search was done at two distinct points. The first term was searched on September, 26, 2008, and on January 16, 2009. The second term was searched on September 29, 2008, and on January 16, 2009.

3.3 YouTube™ by the Numbers

It is possible to gather and analyze a vast amount of quantitative data from YouTube. The purpose here is not to provide an exhaustive analysis but rather to show how statistical methods can be applied to YouTube. The analysis done here clearly shows that we need to move away from the usual statistical realm of means, standard deviations, and t-tests. Instead, it is necessary to think in terms of new types of distributions and use models to provide descriptions of YouTube quantitative data that allow us to see differences and changes in the use of YouTube and hence what its content can tell us about its community.

3.3.1 Gathering Quantitative Data

Like many virtual communities, the YouTube site is awash with information, be it links to other content, statistics about the particular page, or comments by members of the community. Specifically, there are many numbers that are easy to obtain, such as how many times a video has been viewed, viz. view count, the number of comments made on a video, the average rating of a video (one to five stars), and the number of ratings that have been made. With a "scraper" (Cha et al., 2007), it is easy to compile even more numbers based on the complex network of links between YouTube subscribers, the videos, and the commenters. However, as very few papers use or analyze any of these numbers, we focus here on those most easily obtained.

While it is easy to get a view count, even such an apparently straightforward number needs to be questioned. In a search result, each thumbnail for the found videos has next to it a view count along with the time the video was uploaded and its rating. However, on viewing a found video, the view count figure can be somewhat different. For instance, at the time of writing, the search term "Braid" (a recent and highly rated videogame) returned the clip "Braid into gameplay" (http://uk.youtube.com/watch?v=br3Oo1g9nZQ) as the first returned hit. The thumbnail view count was 51,412, but on the page with the video, the view count was 51,562. This is a small discrepancy, but it does beg the question of where these different numbers are coming from. Also, this does not seem to be a difference due to numbers of viewings between getting the search results and hitting the page, as refreshing the search page after clearing the browsing history does not produce the larger figure. On the whole, it seems that search results' view counts are always of the order of a few hundred lower than the figures given on the video page itself.

Additionally, YouTube will happily proclaim a huge number of hits for a given topic. A search for "Shadow of the Colossus" on January 16, 2009, was said to have about 3,310 hits, but going through each page of search results, the results usually peter out after a few hundred, in this case 349. After that, the pages return to the first search results. Of course it could be that this initial hit estimate is wildly inaccurate, but it seems unhelpfully wrong. And as will be seen, it seems likely that there are further videos, but they are simply not being returned.

3.3.2 Looking at View Counts

As YouTube is all about viewing videos, the view count data seems a good place to start to understand the community that uses YouTube, in particular what they view and how that viewing profile changes. The search term "Shadow of the Colossus" was said to have 2,720 hits on September 29, 2008, and 3,310 on January 16, 2009; but of these, 366 were accessible on the first occasion and only 349 on the second. Where the previously accessible videos had gone can only be a matter for speculation.

When videos are ranked by view count, it is clear that the view count drops off extremely rapidly with rank. This is shown in Figure 3.1a, and a good example

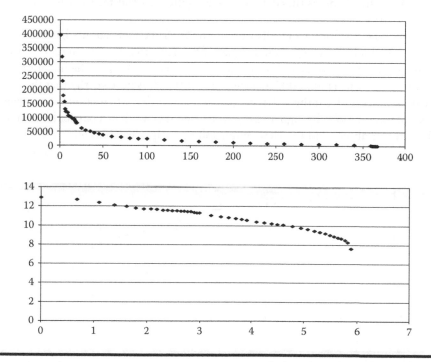

Figure 3.1 (a) A plot of view count against ranking for videos retrieved for "Shadow of the Colossus" on September 26, 2008. (b) A log-log plot of the same data.

of this is the fact that the most viewed video was viewed just over 395,000 times, whereas the twentieth most viewed was seen fewer than 80,000 times. With this sort of distribution, it is usual to take a log-log plot of view count against rank. This greatly reduces the numbers, particularly large numbers; but also a commonly found distribution is the Zipf distribution, where the frequency or size of an object is in a power law relationship with its ranks. So in this case, a Zipf distribution would be that the view count, *v*, is proportional to some power of the ranking, *k*. This is expressed as

$$v = a.k^{-s}$$

This distribution has been found to occur in many situations from word frequency, city sizes (Zipf, 1949), Web site popularity, and Web site links (Adamic and Huberman, 2002).

On a log-log plot, the Zipf distribution produces a straight line with gradient −*s*; so it is particularly easy to identify. However, as can be seen in Figure 3.1b, the log-log plot of view count against rank, though initially linear is clearly deviating from linear at the point at which the data run out.

This "knee" in the data is more clearly seen in the smaller sample retrieved for "Shadow of the Colossus timed attack" (see Figure 3.2) and for the later search on the same term.

This deviation from a Zipf distribution in YouTube view counts has been seen before when looking at very large samples of YouTube that have been scraped from the site by following links within the site, rather than looking at the hits for a particular search (Cha et al., 2007; Cheng et al., 2007). It is noteworthy that the same feature appears in this quite different context but also that when samples are taken a different way, namely through looking at the videos viewed on a particular campus, this feature is absent and the view counts follow a Zipf distribution.

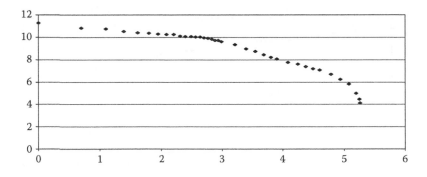

Figure 3.2 A log-log plot of view count against rankings for videos for "Shadow of the Colossus Timed Attack" on September 29, 2008.

3.3.3 From Parameters to Models

Statistical analysis of course goes beyond simply presenting a description of the data collected (though this is of course useful.) Rather, with the use of inferential techniques, it is possible to quantify the nature of the underlying populations and so make predictions about future behavior or other samples. Thus these techniques allow a researcher to ask questions about whether particular samples are different from other samples or differ in behavior from underlying populations from which the samples are drawn. For example, do the "timed attack" view counts reflect a different usage of those videos from the more general "Shadow of the Colossus" videos? Or knowing that this is a dynamic resource, is it possible to discern the growth of the numbers of videos on these topics from the instantaneous snapshots taken?

The key to answering these questions is being able to provide a generic account of how the underlying distributions behave. In the case of parametric statistics, the typical tests, namely t-tests and ANOVAs (analyses of variance), use knowledge of the underlying normal distribution to see whether apparent differences in sample means can be attributed to chance or are due to some underlying influence affecting the samples differently. This equates to determining whether the parameters of the normal distributions underlying the samples, viz. the means, actually differ. The data gathered on view counts though are clearly not normally distributed, so these approaches will not work.

What is needed to achieve the same ability is to be able to determine the parameters of the underlying populations from which the samples are drawn. Had the data followed a straight line in the log-log plot, it would be natural to assume that the data followed a Zipf distribution that is entirely characterized by the exponent s in the equation above (a is also a parameter but basically reflects the number of people viewing the highest ranked item so is not as important in determining the underlying distribution.) Moreover, it is well known that the Zipf distribution arises naturally in a process called the Yule-Simon process (Wikipedia, 2009c). This process can be abstracted to a situation where there are urns with balls in them (a favorite scenario of mathematicians), and balls are added to the urns or more urns are added to the collection. To produce a Zipf distribution, there is a large number of turns. On each turn either a new ball is added to a chosen urn or a new urn with one ball in it is added. In the first case, the urn is chosen at random but the probability of a particular urn being chosen is in proportion to the number of balls already in the urn. This is characterized as "the rich get richer." Whether a ball is added to an existing urn or a new urn is added is also a random process where the probability of adding a new urn is usually quite small. After many turns, the numbers of balls in the urns follow a Zipf distribution for the majority of the urns, in particular in the lower ranking urns.

Thus, knowing what the distribution ought to be not only gives the power to determine the parameters of the distribution but also provides a process by which that distribution arose. Clearly then, the Yule-Simon process is *not* driving the viewing of YouTube videos despite the intuitive appeal of the "rich get richer" scenario together with a steady, probabilistic rate of growth, at least for specific search results such as "Shadow of the Colossus."

There are of course other distributions that are able to account for the knee observed here. Cheng et al. (2007) proposed Weibull and gamma distributions as possible, though still imprecise better fits for the data. These correspond to a stretched exponential distribution (Guo et al., 2008) or a summed exponential distribution. Cha et al. (2007) account for the knee with a Zipf law followed by an exponential cutoff. They offer reasons for this cutoff, but there is no empirical support for these reasons. So in general, there is no explanation of why the processes underlying these distributions account for the observed data. Also, it should be noted that these papers consider large sections of all YouTube content and do not necessarily account for how particular search topics change. For instance, it is apparent that some topics, such as those being considered here, will fade in popularity over time and, therefore, by comparison to other videos. This will affect the overall shape of the view counts away from a Zipf distribution. However, if a person has entered a particular search term, that shows an interest in that topic over and above other topics, and hence although the rate of viewing and the rate of contributing new videos may decrease with time, this does not account for the deviations from a Zipf distribution *within* that topic.

Considerations of this sort may also account for Gill et al.'s (2007) findings that viewings of YouTube videos on a particular campus do follow a Zipf law. The data were collected only over eighty-five days, and the most popular video was only viewed around a thousand times, suggesting that at most one thousand users viewed that video. Thus, the opportunity for changes in popularity was less when compared with YouTube as a whole or for a topic that has been present on YouTube for at least two or three years. Also, with a campus offering the opportunity for vectors of communication between viewers (other than YouTube itself) combined with relatively small numbers of viewers, it is possible that the Yule-Simon process is a good underlying process for the viewing of videos in this context.

Thus, although it is nice to be able to put a name to an underlying distribution for view counts, there is always the risk that this is just fitting the data and there is no good reason for the distribution to be one shape rather than another similar one. We would hold that it is much more useful if the underlying distribution can be supported by a model or process that indicates why this viewing behavior arose, and so we turn to this next.

3.3.4 Modeling View Counts

The Yule-Simon process clearly has key aspects that seem suitable for YouTube viewing behavior. Search results in YouTube are weighted toward highly viewed videos and also, using sorting by view count, viewers can start with those most highly viewed very easily. This promotes a rich get richer attitude of the Yule-Simon process. Also, it is clear that videos are added to cover current topics, and this matches the growth aspect of the Yule-Simon process; however, clearly the two features are not as balanced in view counts as they are in the ideal Yule-Simon process.

Thus, we assume that the underlying general behavior of the rich getting richer does generally apply when it comes to viewing YouTube data but that the growth of videos on a given topic is not in the right balance for the Yule-Simon process. Java was used to implement a model of a collection of YouTube videos that were viewed on a rich get richer basis. That is, there was always the possibility of any of the videos being viewed, but that possibility was in proportion to how many times it had already been viewed. The model was used to vary the way in which the underlying collection grew to see which variations produced the observed "knee" in the view counts. Parameters that could be varied within a given model were the number of videos in the collection, the rate of growth, and the proportion in which "rich" videos got richer, that is, whether a video getting a viewing resulted in only one further view or multiple further views before another video was viewed. We call this parameter the "view increment."

The extreme case would be as in a very old topic (in YouTube terms) where a topic is still being extensively viewed but there are no new videos. In reality, it is not clear that there are any such topics. There are certainly old videos, such as the electric guitar performance of Pachelbel's Canon in D (http://www.youtube.com/watch?v=QjA5faZF1A8), that are still being viewed and commented on despite having been uploaded since more or less the beginning of YouTube. However, a search for this topic shows that videos are still being uploaded that match the search term.

When modeling this unrealistic extreme, for a range of video collection sizes and view increments, the resulting view counts took on a generally uniform shape, as seen, for example, in Figure 3.3. However, what is noticeable is that the overall

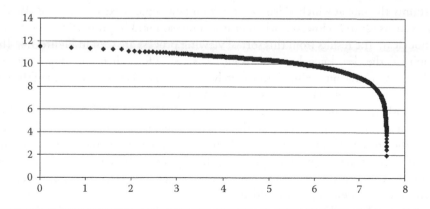

Figure 3.3   A log-log plot of view counts ranking for a model of 2,000 videos on a rich get richer basis but without any growth in the number of videos.

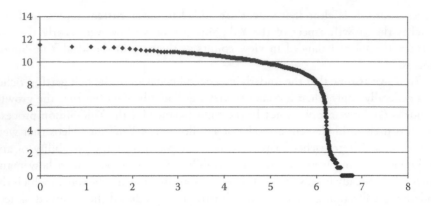

Figure 3.4 Log-log plot of view count against ranking for a model with initially 500 videos and a growth rate of 5 new videos every 100,000 viewings.

shape of the curve is quite flat in the log-log plot until the point at which view counts drop off. This is quite encouraging, as it produces the distinctive knee seen in the real data but is not realistic in terms of how real videos are viewed.

With growth in videos as modeled by the Yule-Simon process (the other extreme), our model produces the expected straight line in the log-log plot. Thus something between regular growth and no growth must be happening. Exactly what is happening can be found only by some form of extensive and detailed longitudinal observation, but using these simplistic models it is possible to see how different patterns of growth might affect overall view counts.

A simple model would be to imagine that the initial growth is initially so rapid that, in effect, there is a whole set of videos available to watch from the outset but that after that growth is only extremely slow. It is not clear that this is realistic, but certainly the rate at which videos appear for a new topic is extremely rapid (Blythe and Cairns, 2009). However, whether it is quite so rapid in proportion to viewing is not clear. The results from this sort of viewing process are seen in Figure 3.4. The graph has the clear knee, but there is also a "scree" effect where the drop off in the tail begins to level out again. Unfortunately these data on how the very end of the tail behaves are precisely what is missing from YouTube.

Another, more plausible model is that initially there is a steady growth in videos but that at some point this tails off though viewing still continues at a modest rate. How the growth of videos on a topic changes is entirely unknown, and so it would be pointless to try to provide a detailed description of how the growth tails off. Thus, we have used a very crude model where after the video collection reaches a certain size, no new videos are added.

For a video collection whose growth is capped at two thousand videos and the initial growth rate is high, the model does produce view counts very similar to those seen in YouTube. There is a pronounced knee in the log-log plot, and by

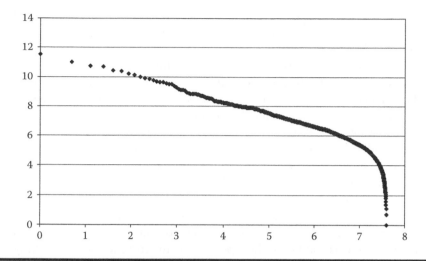

Figure 3.5   Log-log plot of view counts against ranking for a model where the growth rate is one new video for every five viewings until there are 2,000 videos.

varying the initial growth rate, it is possible to vary the plot from anything between a straight line to that seen in Figure 3.5. Of course, the initial growth rate is very high, one new video for every five viewings on average, but the capping suddenly stops the growth, and it begins to develop the knee shape. However, for a real topic like iPhone 3G, some sort of explosive initial growth seems reasonable while people compete to be the first to cover the new product in some distinctive way. Also, it seems reasonable that this initial excitement to upload videos tails off. Our model reflects this without knowing the details of the tailing off.

3.3.5 Using the Models

The different models of viewing based on a rich get richer principle but with varying growth rates seem able to capture the features observed in the quantitative data gathered from YouTube. Thus it seems that a modified Yule-Simon process does adequately capture how videos are viewed on YouTube. However, without good empirical knowledge of the actual growth rate and how it slows over time, the models are at best qualitative.

In addition, the models differ in how the tail of the distribution behaves, which has already been observed as a crucial feature elsewhere (Cha et al., 2007); but it is precisely the behavior of the tail that is missing from the data provided by YouTube.

Thus, the first step in producing a good quantitative analysis of YouTube numbers is to gain better empirical knowledge of viewing behavior. This would require a large-scale analysis of topics over a long period to provide good estimates of the variety of growth rates and their tailing off. With such information, it would be straightforward to produce models like those used above to simulate the viewing behavior and so allow deeper analysis. In fact, it may be that a good knowledge of the growth rates would lead to an analytic definition of the view count distributions.

Once we had good models, we would be in a position to ask key questions, such, as how does the growth rate vary between different topics? From a snapshot of view counts, would it be possible to deduce some of the history of how videos had been added to the topic? And, how representative are subtopics of the overall topic? Answers to these would provide useful insights into how YouTube reflects the changing interests of its online community. Achieving these answers may be a mix of traditional statistical analysis and further modeling using Monte Carlo techniques (simulating samples based on the models.)

Of course, these models might apply only to YouTube. But the general principles seen are likely to be more generally valuable in other contexts. In the first place, good data gathering needs to be done to see what the numbers have to say about a virtual community then using models to understand what processes are driving the numbers and hence allow researchers to return to the data seeking the range and variation possible within the processes. Once a deep understanding of the underlying processes is gained through modeling, it should be possible to ask questions about what is represented in the online community and how it changes and develops over time.

3.4 Content Analysis

Content analysis is frequently used in studies of mass media. It is a procedure for studying textual data, where text is understood to mean any media (e.g., film, newspaper articles, and advertisements.) Items are coded and counted to indicate patterns and trends in data sets. A simple example of this would be counting the number of column inches in newspapers for particular stories. The method was most famously used by Noam Chomsky, who has shown that Western media consistently devote more column inches to the deaths of Israelis than to the deaths of Palestinians. Although such results may be controversial, the method is entirely empirical (e.g., Herman and Chomsky, 2008). Al Gore also used the method to great effect in his documentary *An Inconvenient Truth* when he compared the number of articles in the popular press that doubt the cause of global warming with those that doubt it in peer-reviewed articles. Whereas more than half of the popular press articles doubted whether human activity was causing global warming, the percentage in peer-reviewed scientific articles was zero. Again the findings may be controversial, but the method is straightforwardly yielding quantitative results that can be compared over time. Content analysis can be applied to both quantitative and qualitative studies.

Coding data by topic is not always entirely straightforward, and there may be debate over whether particular items of content fit particular categories. For example, a content analysis might be performed on sexist imagery in advertising (Herman and Chomsky, 2008). There may be some debate as to what constitutes a

Figure 3.6   A traditionally sexist advertisement.

sexist image. Images from the 1950s such as that of a husband spanking his wife for using the wrong sort of coffee (see Figure 3.6) are, to current enlightened eyes, quite clearly sexist.

Advertising standards bodies would not allow such images today and advertisers have resorted to more subtle appeals. The image in Figure 3.7 is cited as an example of contemporary sexism in advertising, but it is not quite as crude as the previous example. Regardless of whether the image is degrading to either sex, it clearly makes an appeal at the level of gender: Women like bingo don't they lads? Content analysis can proceed so long as a broad definition of the categories to be used is in place. For this reason content analysis is most frequently used in areas that are comparatively well theorized and studied. The method is very often applied in studies of the mass media where such categories as genre are well defined and there is a large body of work on which to draw.

Although YouTube is still very young, the formats of many of the videos that are shown there are very much older, and lots of the work in media theory can be drawn on to understand it (e.g. Mayring, 2004). This does not apply to the more interactive aspects of the site such as the comments, as later sections will demonstrate.

Figure 3.7 A more modern take on the sexist advertisement.

In order to get a feeling for returns on the search "Shadow of the Colossus," a content analysis was performed on the first one hundred most frequently viewed videos. The search was made on February 5, 2009, and there were approximately 674 results. Table 3.1 and Figure 3.8 summarize the results.

By far the most frequent category of video was the demonstration. Here users recorded themselves playing the game either directly from the game console or by pointing a camera at the TV screen. As *Shadow of the Colossus* consists, essentially, of sixteen boss levels interspersed with scenic journeys, each of the battles lend themselves to a mini-movie format. Each of the sixteen colossus monsters are different, and the chief interest of the videos is perhaps the spectacle of the new monster as it is revealed and then defeated. These were very popular videos: At the time of writing, the "speedrun on colossus 13" had been viewed 198,095 times. The comments to such videos indicate that they are valued not only because they show off the creatures, some of which can only be seen after many hours of dedicated game play, but also because they offer tips to other players on how to defeat them. There is both instruction and celebration in these posts.

The next most frequent return was the movie clip. This is slightly different from the demo in that it does not necessarily feature actual game play but rather clips from the noninteractive movies, which players can watch upon reaching certain levels. Popular videos in this category featured the final end sequences of the game that contain plot spoilers. The "secrets" category refers to secret levels, such as the way to the secret garden. The game play, movies, and secrets are all in a sense demonstrations, and they are separated here only to show additional detail. The three reviews can also be considered forms of demonstration although they are more explicitly evaluative. The four Sony advertisements might also be considered a variety of demo albeit a heavily edited demonstration geared to make a particular kind

Table 3.1 Categorization of Videos Returned for the First One Hundred Hits on the Search Term "Shadow of the Colossus"

Demonstration	41
Movie clip	9
Secrets	8
Mash up	7
Soundtrack	7
Imitation	6
Little Big Planet	6
Advert	4
Glitch	4
Parody	4
Review	3
Mashinima	1

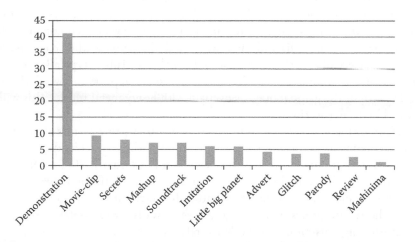

Figure 3.8 Bar chart of the categories of videos returned for the first 100 hits on the search term "Shadow of the Colossus."

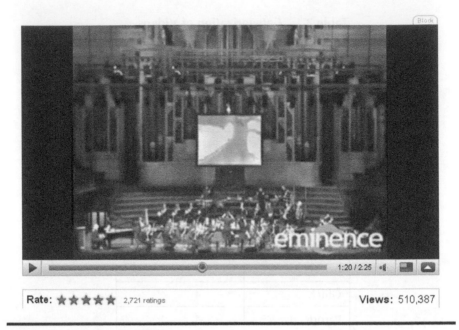

Figure 3.9   A still from "Shadow of the Colossus — Eminence Orchestra."

of impression. Taken together, the demonstrations, movie clips, secrets, and reviews account for more than 60% of the top one hundred most viewed returns.

The content in the remaining 40% of the most viewed videos takes the form of slightly less familiar media. "Mash ups" and "Machinima" are relatively new forms and perhaps the most original genres to emerge in user-generated content. Mash ups take two existing media and combine or mash them up together, e.g., a piece of music and a piece of film (Bardzell, 2007). Here the YouTubers typically combined footage from the game with music. *Carmina Burana* appears in two such mash ups. In the rest, the game footage is overlaid with heavy metal music. This will be returned to in later sections.

Machinima is a form of video production that uses game avatars as actors and game environments as locations. Players manipulate the characters directly in the game or in postproduction to tell their own stories (Bardzell, 2007). Often such videos are comic but not always. The video "Agro-vation!! (funny)" begins with a series of game play clips in which the central character attempts to get his horse to lead the horse into places it cannot go (up sheer walls for example), and subtitles present growing frustration as the horse refuses to cooperate. Eventually a sub-title appears announcing that this has to stop and the character proceeds to kill the horse in "entertaining" ways (e.g., running it over a cliff.) The video ends by exploiting a glitch in the game that the "glitch" videos also note. When the horse disappears, he can be summoned back; by killing the horse in a particular location, players exploiting the glitch can make the horse appear in midair. This is the climax of this not terribly inspired piece of machinima.

Similar posts to this are imitations and parodies of the game. The imitations were typically made with very basic animation so that a small spot appears to attack a stick man. A live action parody features a young man with an action figure attached to his foot. He does not notice it as he walks around the house until he brushes his teeth, plucks it from the top of his head, and throws it away. One well-produced imitation features animation in a manga style that is a straight homage rather than parody. Some of the *Little Big Planet*™ posts use this PS3 game creation environment to make mini-colossus games; these appear as part homage and part parody. Others appear completely irrelevant, and it is hard to see why they are returned for this search.

The mash ups, machinima, imitation, parody, and glitch videos all playfully undermine the game. The parodies directly satirize the somewhat grandiose tone by juxtaposing the stirring music with knocked-down graphics, line drawings, action figures, and *Little Big Planet* dinosaurs. The glitch category explicitly draws attention to game mechanics, and these videos achieve their effects by pointing out the limits of the game architecture—the callback feature of the horse making it float in particular environments. All of these posts are in a sense outside the game and the game world. They step back from the Colossus universe and comment on it in imitations that to one degree or another either are parodic or point up its technical or graphical shortcomings.

The posts in the soundtrack category are for the most part entirely celebratory. They feature moments from the game in which new themes in the music are introduced: "The Opened Way," "Demise of the Ritual," "A Violent Encounter," and so on. The music to this game is one of its most popular features, indeed the most frequently viewed post for the entire search is an Eminence Orchestra performance of the "The Opened Way." Two of these posts feature users playing the piece themselves on the piano.

To understand why this post is so popular, the comments on this video will be explored in the next section. As YouTube is only three years old and the ability to comment not only on video posts of this kind but also on the comments is entirely new, a content analysis would not be appropriate. "Grounded Theory" is a form of qualitative data analysis and one of the most frequently used. It begins not with preexisting categories but with the data themselves.

3.5 Grounded Theory

Grounded theory was developed in the late 1960s by Barney Glaser and Anselm Strauss. The two originators of the approach subsequently took it forward in different ways. Strauss wrote a book with Juliet Corbin that presented itself as a guidebook on qualitative data analysis (Strauss and Corbin, 1990). This set out a quite formal approach delineating three stages in the process of conducting grounded theory: open coding, axial coding, and selective coding.

Open coding describes the process whereby complex data such as interview transcripts or field note observations are summarized with single terms or two- or three-word phrases. Depending on the size of the data set, there may be a great many such open codes. In the next stage of the process, "axial coding" relationships between the open codes are established. Similar or related codes are grouped together into a smaller number of larger metacodes. Relationships between the resulting terms are then considered, as it were, connecting the axes. In practical terms, the researcher now has a set of data that has been organized into overarching themes. At this point "selective coding" begins. This is essentially the creation of narrative introducing the axial codes and illustrating them with quotes. In the process of describing the codes and the relationships among them, theory is said to emerge.

The production of this book resulted in an academic spat with Glaser (e.g., Glaser, 1998). Despite their emphasis on technical precision, Strauss and Corbin are unclear as to how the categories generated during open coding and linked in axial coding create theory. Glaser argued that their positivist stance was inappropriate and emphasized instead the individual researcher's creativity. Glaser's coding procedure begins with open coding but does not include an axial stage. Rather the open coding is followed by theoretical coding. The researcher is encouraged to generate theory throughout the encounter with data, making detailed notes and memos on the fly. Further, the researcher is discouraged from conducting a literature review related to the topic being studied as it might unduly influence the analysis. They are also discouraged from taping interviews or talking about the grounded theory until it is complete, praise or criticism being judged as unhelpful and distracting (Wikipedia, 2009a).

Although there are differences in orientation between the two schools, they have in common a data-led approach to theory. Despite disagreements between Glaser and Strauss, grounded theory has become one of the most frequently practiced forms of qualitative data analysis. Practitioners of grounded theory often use the term "theory" very loosely. "Theory" here may refer merely to a broad description or set of categories rather than a fully worked predictive schema (Charmaz, 2006). The following section, then, is a grounded theory analysis of posts on the Eminence Orchestra video.

3.5.1 Eminence Orchestra

At the time of writing, "Shadow of the Colossus—Eminence Orchestra" had been viewed 510,387 times. The orchestra members are dressed in black although they are not wearing complete eveningwear. The conductor wears a suit but no tie. It is a formal performance but not entirely traditional. Above the orchestra is a large video screen showing footage from the game (see Figure 3.9). As particular instruments are featured in the music, the camera focuses in on the relevant section of the orchestra. As the piece builds to its climax, a snare drum beats out a militaristic tattoo, and the camera turns to the percussionist playing it. Although many of the most popular video posts here showcase the music, this is unusual in that it is a live performance. The "liveness" is emphasized in the attention to the individuals playing particular parts of the piece.

The post not only had been viewed over half a million times, it also was very highly rated, with an average of five out of five stars from 2,721 ratings. There were also 1,481 comments posted on the piece.

The "all comments" tab was selected, and the first 250 posts were coded by copying the comment into one column of Microsoft Excel and writing an open code that summarized it in another. The data completely saturated long before the 250th comment was reached, i.e., the addition of new data did not require the creation of a new code, it just slotted into one of those already created. The open codes and the frequency with which they were used are shown in Table 3.2 and Figure 3.10.

Table 3.2 Open Code Frequencies for Comments on the "Shadow of the Colossus—Eminence Orchestra" Video

Awe	51
Music appreciation	28
Game appreciation	26
Soundtrack appreciation	26
Liveness	24
Music availability	23
Comparison	15
Affect	13
Criticism	13
Nostalgia for game	10
Eminence information	8
Game experiences	6
Learning the music	3
Flames	2
Wishing it real	2

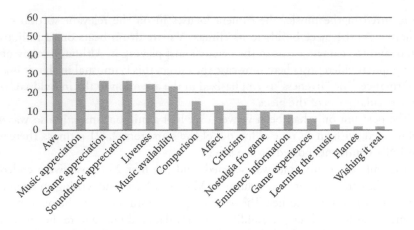

Figure 3.10 Bar chart of the open code frequencies for comments on the "Shadow of the Colossus — Eminence Orchestra" video.

Glaser would immediately begin to theorize about these codes. Indeed it is difficult not to. For illustrative purposes, the next section considers how these codes might be further categorized into axial codes.

3.5.2 Axial Coding

A Straussian grounded theory would take these codes and attempt to organize them. An obvious point of similarity for some of these codes is appreciation. One overarching or "axial" code might then be "**appreciation**," and this would include music appreciation, game appreciation, and soundtrack appreciation. "Awe" refers to general comments such as "awesome!" or "epic!" and other such superlatives. Similarly, "affect" refers to strongly expressed feelings in response to the post, so these might be put together under an axial code like "**emotion**." Nostalgia for the game might even be included here. Comparison to other soundtracks and media might be grouped with criticism. "Flames" were angry responses to criticism and so could be grouped here under a heading like "**evaluation**." "Liveness" refers to comments about the fact that the music is being performed live by an orchestra, so this might be included here as well. "Music availability" included requests and replies about where to download mp3s or buy CDs. "Eminence information" referred to posts about upcoming concert gigs or previous performances. These categories might then be grouped as "further information" and considered as a subset of appreciation Game experiences and nostalgia for the game might be grouped as "**Remembrance**." "Wishing it real" contained atypical, slightly odd posts in which the users wished that the colossi were real and imagined them walking down their streets. It does not quite fit with remembrance, but they are a form of imaginative response to having played the game so might sit better here than in any of the other categories. The fifteen "open" codes would then become four axial codes: appreciation, emotion, evaluation, and remembrance.

At this point Strauss and Corbin would suggest that there may be relationships between the axial codes. Qualitative data analysis software packages such as NVivo and HyperResearch offer a number of tools for making graphs that help identify such relationships. Of course, there is any number of ways to connect such abstract concepts, and this is the point at which many students of grounded theory get stuck. Still, let's give it a go anyway, shall we? Hmm. So we could link them up like this:

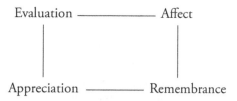

It is important to remember that this kind of representation is little more than a visual aid. It does not represent any defining analysis in the way that a Venn diagram might. It is not claiming a relationship between the categories, one leading to another, nor is it suggesting polarity. It does, however, indicate proximity. Evaluation and appreciation are along the left-hand side of the diagram. They are similar responses. Some sort of evaluation is implicit in any sort of appreciation. Indeed appreciation could be subsumed by the still larger category of judgment. Remembrance calls on previous experiences of game play. Emotional responses are also the result of experience. And again it would be possible to fold these codes into one still larger one, experience. These axes might be presented like this:

Judgment ———— Experience

Again, the diagram, if it can be thought of as such, is merely a visual, aid and in some ways it is quite unhelpful. It suggests a polarity between the two categories although there is a reciprocal relationship between the two. Indeed this relationship can be described as a dialogue.

And here something like a theory is beginning to emerge. The comments board is a space where these gamers can reflect on their experiences, not solely of the music but also of the game that informs evaluations of that experience. This is more descriptive than predictive, but it does provide a structure through which to present the data and conduct further analysis.

Tree diagrams are frequently used to express the kinds of relationships between codes explored above. A possible tree diagram for this set of codes is shown in Figure 3.11.

Again, this is more of a visual aid than any scientific statement of cause and effect. It does, however, indicate a general shape that can be explored in more detailed selections of the data.

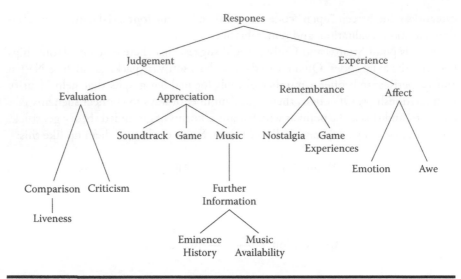

Figure 3.11   A tree diagram of the concepts in the grounded theory.

3.5.3 Selective Coding

Broadly, the codes fall into two categories: judgment and experience. The following sections will explore them in more depth. Strauss would call this "selective coding." Glaser would describe it as "theory building."

3.5.3.1 Judgment

By far the most frequent judgment was positive appreciation. A large number of these responses focused solely on the music, for example, "*I love the part with the drum in it.*" The posts coded as "soundtrack appreciation" rather focused on the music as an accompaniment to the game: "*Awesome soundtrack. Loved this song. It got me pumped whenever I was close to taking down a colossus. This game rocks!!*" A number of these responses indicated that the music was one of the most powerful aspects of the game; indeed, it was crucial to the game play itself. Closely related to this kind of response were those that focused solely on the game. A number of the responses claimed that this was the best game of all time. One user echoing the comic book guy in *The Simpsons* expressed this most succinctly in the three-word post: "*Best game ever.*" Occasionally this led to minor flame wars where some suggested that *Prisoner of Zelda* deserved this accolade more, but the majority of posts disagreed: *Shadow of the Colossus* or "SotC" was the best game of all time. These superlatives are interesting because they lay claim to lasting significance in media that they are well aware is ephemeral and fast moving. Many of the claims were that this was the best game "to date" anticipating the possibility of better games to come but maintaining that, nevertheless, this was a significant historical achievement.

The enthusiasm for the game, the soundtrack, and this particular video were frequently expressed in requests for further information about the Eminence Orchestra, where they play, what the conductor's name is, and so on and also the availability of the soundtrack on mp3 or CD: *"Anybody know where i can get *coughs* get the soundtrack?"* The cough here is interesting: *Can I buy the CD? If not, can I download it somewhere?* The game is over but the soundtrack can still be enjoyed and live performances attended. Again, there may be an attempt here to fix the fleeting ephemeral quality of games.

A number of the responses expressed appreciation of the music with either a comparison to the soundtracks of other games, which were generally nowhere near as good or, more interestingly, comparisons to other forms of music: *"Love that piece, my favorite aside from Moonlight Sonata."* The comparison to classical music is interesting because it equates a traditionally high art form with a popular medium. The formulation is even more explicit in this post: *"This must be the best video game music ever. I wish my orchestra would play this. After all, we are teenagers, it would be perfect. But noooo, we play Dvorak. Oh wait, I like Dvorak!"* Although the music is orchestral, it is not necessarily "classical." Some players refuse the distinction: *"That song is epic, me being a fan of both classical music and the game. The classical music just makes the game even better!"* The comparisons to Beethoven and Dvorak are at some level claims to status. Like the superlative accolades "best game ever," there are claims here for the intrinsic value of the game. To what sociologists like Pierre Bourdieu might call "cultural capital," such posts might be paraphrased as "I am not just wasting my time here, I'm appreciating the best music that the culture has to offer as well as the best game." In other words, these posts are defenses of taste.

A small number of posts seek to demonstrate very refined and subtle apprecia-tions of the music in pointing out minor flaws and imperfections: *"Ouch someone's out of tune in the strings and the snare was a bit too loud … nevertheless, magnificent performance."* On occasion this resulted in hostile responses: *"I'd like to see YOU do better"* if not outright flames. In one exchange, a license to speak is demanded quite explicitly: *"You better be in an orchestra or something."* Again, the comments are not just statements about the game or the music but statements about who the players think they are. In taking the game and the music seriously, they demand that their own interests and tastes be taken seriously.

This is most obvious in the comments coded under "liveness." A large number of the comments focused on the fact that the music is being played by an orchestra. This orchestra specializes in playing game show themes, and there are other clips on YouTube of them playing the theme to *Super Mario*. There may be an extent, then, to which such performances are ironic, a set of instruments associated with the most serious "high" art performing themes from "low" art games. However, the performance is not enjoyed in an ironic way: *"Yeah … well it's good to see an orchestra put the effort of playing this piece."* Whether the performers' intent is ironic or not, the performance itself is proof of time and effort devoted to the piece: *"I've always loved the SoC soundtrack. It always gives me shivers to see this kind of music, specifically game music, preformed live. To see and hear the sounds come from human fingers and tangible instruments rather than trying to imagine them as you listen always amazes me."* That this lyrical comment was answered with the word *"nerd"* suggests the enthusiasm is not necessarily universal. The thrill of a live event was also indicated in requests for further information about where the Eminence Orchestra might be playing. Enjoyment and involvement in the music on the part of these musicians is also seized on as proof of their commitment to the music. Here the conductor's seriousness is explicitly noted: *"The music conductor's demure is most confident ... his bow is so deep! Not only this piece could be spoken of as historically stunning, i was truly amazed at the conductor's outstanding confidence in himself, some people just, never fail to amaze me...sodesune!!"* Similarly, the intensity of the violin player was much admired: *"The violinist to the front left is really into this song. He plays it with a passion. I believe he is also the creator of Eminence if I am not mistaken. Notice the conductor shaking his hand at the beginning."* And more demotically: *"The guy that bangs his head is fucking sick heheehehe."*

There is a sense then in which the performance of the piece by a live orchestra validates the intense feelings that the players have about the game. Although their responses to the piece are clearly profound, there is unease and insecurity about whether it should be. This is perhaps an indication of the trajectory that games have taken. From its relative youth in terms of its technological development and traditional market, the medium is beginning to take itself more seriously. Young gamers play on into adulthood, and the *Shadow of the Colossus* is an ambitious game. Its opening sequences are self-consciously filmic, echoing artistic and cinematic conventions.

3.5.3.2 Experience

Many of the responses to this video indicated that quite powerful emotions had been stirred. This was often expressed as a desire to go and play the game all over again. Occasionally this sentiment was expressed humorously: *"Feeling the need to kill something big and lumbering about now ..."* In other posts, the tone was more nostalgic: *"Brings back a ton of memories beating this game."* This nostalgia occasionally became brief exchanges about previous experiences with the game. Some posts enquire, for instance, if other players managed to reach the secret garden.

Other posts were more directly emotional; those labeled as "affect" were the strongest expressions of emotion: *"Oh my god. A lump formed in my throat and my eyes teared up when I heard this open up. Incredible."* A surprising number refer to tears: *"Tear slides down cheek."* It is possible that such posts are ironic, but some of these players were clearly very moved by the game: *"When i listen to that song i feel something i never felt before ..."* Occasionally such exclamations end in recommendations: *"Oh my god, ... i have tears in my eyes ... this game was the most intense, and most emotional that i have played in my life. Buy this game,"* as if the emotional outburst must be tempered with utilitarian injunctions to buy: I'm not just expressing myself I'm offering advice here. Similarly the reference to tears are tempered by slightly macho cursing: "*Holy crap it made me cry.*" Occasionally the posts express not only deep attachment to the game but also puzzlement over why it affects them so deeply: "*Everything about this game makes me cry with joy, why?*" Similarly, "*Dude i know what u mean it's like so weird. This symphony has like a spell that it casts on u. lol.*" The term "dude" is interesting as it indicates an assumption that the addressee is male. Indeed there appears to be something of a conflict in some of the posts between the intense emotion being expressed and notions of traditional masculine traits like toughness One post stating, "*I must not cry*" was answered with, "*Shut up.*" Admission of intense emotion and bewilderment at that emotion appear to be disturbing some of these players: "*I don't know what happens to me when i listen to this*" The colon bracket indicating a frown suggests that this is not frivolous or fatuous. The players do not understand their own reactions.

By far the most frequent code for responses to this video was "awe." This is in part a term taken from the data itself and so a particularly good example of how a grounded theory analysis differs from a content analysis. An astonishing number of these posts include the word "*awesome*" Of the fifty-one posts coded as "awe," twenty of them featured the word "awesome" itself; sometimes the post contained only that word. Others, although not containing the word, indicated the emotion strongly: "*Wow … just … wow*" or "*Whoa, how incredible is that?*" or "*Breathtaking. I love it.*" The word "awesome" is of course currently used colloquially as a general term of approbation. However, it also continues to connote its dictionary definition of holy terror. One post fits this almost perfectly: " *… mein god.*" For many of these players their awe left them almost speechless: "*Wow, I'm just speechless … wonderful …*" or capable of saying only single words: "*wow*" or "*Wow*" or "*WOW*" or "*just wow.*" For some, their enthusiasm turned their words into long, drawn out vowels sounds of appreciation "*waaaaaaaaaaa iiiiiiiiiiiii excelent iiiiiii*" and "*WAAAAAAAAAAAAAAAAAAAAAAAAAAAAAAAA AAAAAAAAAAAA my favorite song of SotC T_T*"

It is interesting to note that a great many of the posts expressing general approbation do so in violent terms: "*two words………KICK ASS!!!*" "Great," "awesome," "rocks"—this is the language of shock and awe. It is as if their profound emotions can be expressed only in terms of violence.

Clearly, a large number of these players are teenagers. Their user names indicate a range of interesting connotations that would make up an interesting study in itself.

Fsmetal
Fantasygurl14u
Unwingedangel
Tromboneprincess
HyperBebe123
SakuraBubbles
LadyAssasin

Thewatermargin
Mrbigtime81
Boywithhorns
Shanksxy
POWERHOAX
Awesomeuserdude
MortalSyn

Although some of these names strongly suggest gender—"awesomeuserdude"—most are ambiguous, referencing TV shows, manga, and fantasy worlds rather than gendered names.

It has long been a commonplace of research in virtual communities that they provide a space for experimentation with different identities (Turkle, 1995). But at play in these posts is a range of tensions around not only gender but also cultural capital. They might be paraphrased as "Is this profoundly emotional response to a game all right. Is it OK for me to be crying over this game? Is it OK for me to feel profoundly moved by this video game theme tune?"

What then might a formulation of this grounded theory look like? Perhaps something along these lines:

The responses boards provide a space to express judgments and experiences of a new cultural form; although this form echoes other older arts it is new and therefore its value is uncertain; the purpose of the posts then is ultimately reassurance about this still evolving, bewitching new media.

Theory would not necessarily usually be spelled out as specifically as this. There is value in ambiguous description. Indeed, and as previously noted, a new set of categorizations can in itself be considered theory without further elaboration.

And that's something like a grounded theory. It's not going to win any Nobel prizes, but there we are. Although Glaser would counsel against it, grounded theories are often supplemented with other theories drawn from disparate areas.

3.6 Critical Theory

Critical theory incorporates a number of theoretical perspectives, such as structuralism, poststructuralism, feminism, Marxism, and psychoanalysis. This pluralistic approach to cultural analysis has grown from several sources. In the early 1950s, Chicago school theorists such as Theodor Adorno developed a mixture of Marxism and phenomenology that sought to analyze and, moreover, change mass culture. In the early 1960s, French theorists such as Roland Barthes and Umberto Eco also turned their attention from "high" to popular culture. Barthes had studied literature and used techniques of close and detailed reading to analyze not poems and novels but "texts" such as films and commercials. In the 1970s, the Birmingham Centre for Contemporary Cultural Studies (BCCCS) was founded by Stuart Hall.

There had been a long history of studying subcultures in anthropology, but this work was primarily a colonial encounter where (typically) Western field-workers studied non-Western cultures, usually with a view to governing them more effectively. In the BCCCS, analytical tools that the West had developed to understand distant countries were turned back onto itself.

Theodor Adorno is perhaps the best-known exponent of critical theory and one of the first to ask what a sociology of music might look like (Adorno, 1962). Writing on the cultural industries of the 1950s, Adorno and Horkheimer (1986) pointed to the similarities between the ways in which leisure and work time were structured and monitored. For these authors, the cultural industries exacerbated the artificial division between enjoyment and the rest of life: "Amusement under late capitalism is the prolongation of work" (Adorno and Horkheimer: 137). Although amusement is sought as an escape from mechanized work, mechanization determines the production of "amusement goods" with the result that leisure experiences are "inevitably afterimages of the work process itself" (Adorno and Horkheimer).

Adorno disliked popular music, seeing it primarily as a standardized musical form. The rigid structure of the pop song with its exactly repeated verses and refrains was a debasement of the musical imaginations not only of the composers but also of the audience. He was similarly disparaging about movie soundtracks:

"Much of the consumers' music anticipates fanfares at victories yet accomplished, along with the applause. The garishly instrumented film titles which so often seem to resemble a barker's spiel: 'look here, everyone! What you will see is as grand, as radiant, as colourful as I am! Be grateful, clap your hands and buy'— these set the pattern of the consumer's music even where the feats proclaimed by the shouting do not follow at all" (Adorno, 1961: 46).

This could almost be a description of "The Opened Way." The content analysis indicated that a number of users had substituted the game theme with Carl Orff's *Carmina Burana*. It is in many respects a similar piece of music. Orff wrote the music for *Carmina Burana* in the 1930s, but the piece echoes older forms of religious and folk music. The lyrics were taken from a Latin manuscript that dated back to the Middle Ages. The piece was echoed in horror films such as *The Omen*, which also employed large choirs singing in Latin to evoke numinous and sinister effects. "The Opened Way" features no choral singing, but the orchestrations of *Carmina Burana* are clearly an influence, as is Wagner's *Ride of the Valkyries*. The piece then draws on a rich tradition that resonates through many media.

The sense of awe in these related media refers to fear, terror, and dread that has religious connotations. Etymologically, this kind of dread is "mingled with veneration, reverential or respectful fear" in the presence of God or, later, nature. Here the music references earlier forms that also seek to inspire awe of either God (as in *Carmina Burana*) or national destiny (Wagner). Here the musical effects (heavy timpani, crashing symbols, swelling string sections punctuated by dramatic horns in chorus reaching its crescendo in a military tattoo) are employed to create a sense of awe, not before God or Nature or Nation but rather of a video game.

The bewilderment of the awed consumers is in this sense understandable. Should I be awed by this game? Should I fear and revere it? Clearly, the game makers would hope so. The injunction here is undoubtedly to buy. And yet, the performance of the theme in the context not of the game but a of a live orchestra concert subtly changes its meanings. Must we dismiss it as low art along with Adorno? Not necessarily.

Marxist analyses of the cultural industries and leisure are, of course, deeply unfashionable and have been criticized for their pessimism and elitism. Empirical studies on the actual uses of cultural products have shown than consumption is not passive: Private and individual meanings are invested in leisure activities despite hegemonic intent (Willis, 1990). We do not watch TV solely because we have become the numbed spectators of our own lives, passively and joylessly consuming spectacles as what Garfinkel called "cultural dopes."

Indeed critical theory is in some respects at odds with grounded theory. Glaser warns researchers not to read related literature. However, many would argue that it is never possible to approach any data set with an entirely theory-free mind. Nobody can unread a book they have read or temporarily abandon all the ways they have previously made sense of the world. To claim ideological neutrality is itself an ideological position. Acknowledging and sourcing theoretical sources indicates that the account is one among many and may be supplemented or overturned by other perspectives. There can be no final or ultimate meaning to any cultural artifact; there are as many readings as there are readers.

3.7 Discussion and Further Work

In all the approaches discussed, it is possible to provide insights into the nature of the YouTube content and the community who uses it and contributes to it. These insights of course differ with different research methods: Statistical approaches show how modeling might provide the ability to infer changes in viewing behavior of a particular topic over time; content analysis breaks down the different sorts of videos to see what they are "about"; grounded theory reveals how people's response to a video reflects attitudes to old and new media; critical theory suggests how the split between low and high art is writ large in a single YouTube video. Each of these insights is interesting and suggests avenues for further fruitful research.

There remains, however, the challenge posed by the very size and rate of growth of YouTube. Already the number of videos found on a search for the Hudson River is well over a thousand—around double what it was when this chapter began. There have been more than 34,000 further viewings of the Eminence Orchestra playing "The Opened Way" since the last search point in January and a further one hundred videos uploaded that match the search "Shadow of the Colossus."

In terms of content analysis, it is unlikely that these changes pose specific problems. We would make a fairly confident claim that there are no really new types of video being uploaded that would defy the existing categorization or inform it in any significant way. But then, the content analysis perhaps had the least interesting message about what YouTube is, as it simply reports on the kinds of things that appear on YouTube without trying to account for them or their popularity with YouTube viewers.

For statistical analysis though, these changes are quite large, and although the overall shape of the statistics may remain the same between quite large gaps in data gathering, there are definitely circumstances in which the changes are substantial and perhaps defy the models that we are proposing here. For example, since writing up our work on iPhone 3G (Blythe and Cairns, 2009), the most popular video is no longer "Will it Blend?" A new video has a lot higher view count. How is this possible in the rich get richer model? Is something else driving viewing figures?

Another challenge for the statistical analysis is the superfluity of data that YouTube offers. Usually statistical methods work with a sample available and infer properties of the underlying (hypothesized) population. With YouTube, the samples available for a given search term are massive proportions of the total hits that YouTube reports—inferential statistics are barely necessary. And more deeply, why would we need to infer anything at all? To find out more, we just look longer. The models provide potential descriptions of viewing processes, but the question then arises: How generalizable are these models to other topics? The *Shadow of the Colossus* is a videogame and therefore likely to be of at best short-term interest. But what about a different topic that might have long-term relevance, such as "chemistry experiment" where what matters is not being up to date but being accurate and informative (and for which there are currently 2,120 hits)? Will that grow and develop in the same way that *Shadow of the Colossus* has? And what about very rapidly changing topics such as news stories that can have multiple evolving perspectives (angles) and recur, thanks to new information, revived relevance, and so on? We believe the modeling approach is promising; but given the size and scale of YouTube, this is currently, and possibly only ever, a belief.

The same problem of general relevance can be applied to the grounded theory approach. The theory produced here is interesting, but it is based only on the comments of one video that was popular for one topic. Also, there is perhaps a more subtle problem in analyzing the most popular video. This may seem reasonable—many people have viewed that video, and therefore the comments are about something of significance. However, the whole statistical modeling approach suggests that the underlying process is random and that whatever gives a video an edge over other videos, the rich get richer process will continue to promote that video over others. That is, the most popular video may just be a quirk of fate.

Additionally, analyzing more comments becomes an endless task. At what point will it no longer be necessary to produce a grounded theory? This is an identical problem to that faced in statistical analysis. When do you stop analyzing and start generalizing? And on what basis?

Critical theory does, therefore, offer a way to slice the Gordian knot of YouTube's size. Though we have analyzed the most popular video and therefore that may be questioned as for grounded theory, this is irrelevant. Critical theory has picked out a cultural artifact, for whatever reason, and presented a reading of that artifact that provokes ideas and provides insight.

This is not to say that critical theory is thus the best method; indeed it is likely to be the most controversial. Rather it is only through a multimethod approach that it will be possible to provide a rich picture of what is represented in YouTube and the community that comprises and consumes it.

References

Adamic, L.A., Huberman, B.A. (2002). Zipf's Law and the Internet. *Glottometrics* 3: 143–150.

Adorno, T., (1962). *Introduction to the Sociology of Music.* New York: Seabury Press.

Adorno, T., Horkheimer, M. (1986). *Dialectic of Enlightenment.* London: Verso.

Bardzell, J. (2007). Creativity in amateur multimedia: Popular culture, critical theory, and HCI. *Human Technology* 3, 1:12–33

Bell, G., Blythe, M., Gaver, W., Sengers, P., Wright, P. (2003). Designing culturally situated products for the home. *Proceedings of ACM Conference on Human Factors in Computing Systems 2003*, New York: ACM Press, 1062–1063.

Bertelsen, O., Pold, S. (2004). Criticism as an approach to interface aesthetics. *Proceedings of NordiCHI 2004*, New York: ACM Press, 23–32

Blythe, M., Bardzell, J., Bardzell, S., Blackwell, A. (2008). Critical issues in interaction design. *Proceedings of BCS Conference on Human-Computer Interaction 2008*, Vol. 2, London: BCS Press, 183–184.

Blythe, M., Cairns, P. (2009). Critical methods and user generated content: The iPhone on YouTube. *Proceedings of ACM Conference on Human Factors in Computing Systems 2009*, New York: ACM Press, 1467–1476.

Cha, M., Kwak, H., Rodriguez, P., Ahn, Y.-Y., Moon, A. (2007). I Tube, You Tube, Everybody Tubes: Analyzing the world's largest user generated content video system. *Proceedings of the Seventh ACM Special Interest Group on Data Communication Conference on Internet Measurement*, New York: ACM Press, 1–14.

Charmaz, K. (2006). *Constructing Grounded Theory.* London: Sage.

Cheng, X., Dale, C., Liu, J. (2009). Understanding the characteristics of Internet short video sharing: YouTube as a case study. To appear in IEEE Transactions on Multimedia.

De Souza, C. (2005). *The Semiotic Engineering of Human-Computer Interaction.* Cambridge, MA: MIT Press.

Gill, P., Arlitt, M., Li, Z., Mahanti, A. (2007). YouTube traffic characterization: A view from the edge. *Proceedings of the Seventh ACM Special Interest Group on Data Communication Conference on Internet Measurement*, New York: ACM Press, 15–28.

Glaser, B., Strauss, A. (1967, 1999). *The Discovery of Grounded Theory: Strategies for Qualitative Research.* Chicago, IL: Aldine Publishing Company.

Guo, L., Tan, E., Chen, S., Xiao, Z., Zhang, X. (2008). The Stretched exponential distribution of Internet media access patterns. *Proceedings of the 27th ACM Symposium on Principles of Distributed Computing*, New York: ACM Press, 283–294.

Herman, S., Chomsky, N. (2008). *Manufacturing Consent: The Political Economy of the Mass Media*. London: Bodley Head.

Juul, J. (2005). *Half Real: Video Games Between Real Rules and Fictional Worlds*. Cambridge, MA: MIT Press.

Mayring, P. (2004). Qualitative Content Analysis. In Flick, U., von Kardorff, E., Steinke, I. eds. *A Companion to Qualitative Research*. London: Sage, 266–269.

Pspimp. (2009). Shadow of the colossus viral campaign 2. *YouTube*. http//www.youtube.com/watch?v=9TPYhZJqydo. Accessed August 12th, 2009.

Sengers, P., Boehner, K., Shay, D., Kaye, J., Reflective Design. (2005). *Critical Computing*. Aarhus, Denmark. 49–58.

Sony Computer Entertainment. (2006). *Shadow of the Colossus, Playstation 2*. London: SCEI.

Strauss, A., Corbin, J. (1990). *Basics of Qualitative Research: Grounded Theory Procedures and Techniques*. London: Sage.

Turkle, S. (1995). *Life on the Screen: Identity in the Age of the Internet*. Phoenix. London.

Wikipedia. (2009a). Grounded theory. http://en.wikipedia.org/wiki/Grounded_theory. Accessed August 12th, 2009.

Wikipedia. (2009b). Shadow of the Colossus. http://en.wikipedia.org/wiki/Shadow_of_the_Colossus. Accessed August 12th, 2009.

Wikipedia. (2009c). Yule-Simon distribution. http://en.wikipedia.org/wiki/Yule%E2%80%93Simon_distribution. Accessed August 12th, 2009.

Willis, Paul (1990). *Common culture: symbolic work at play in the everyday cultures of the young*. Milton Keyness, Open University Press.

Zipf, G. (1949). *Human Behavior and the Principle of Least Effort*. Cambridge, MA: Addition-Wesley.

APPLICATION AREAS

Chapter 4

E-Learning Communities

Andrew Laghos

Contents

4.1　Introduction

The focus of this chapter are the concepts of e-learning and e-learning communities and relevant research in these areas. This is an important study area because an understanding of how students communicate and the characteristics of online learning communities are necessary in order to be able to accurately analyze them.

An online learning community can be defined as "a group of people who communicate with each other across the Internet (or sometimes by intranet) to share information, learn more about a topic, or work on a project of mutual interest" (Porter, 2004, p. 193).

The importance of online learning communities is reflected in a subsequent increase in support between peers, higher retention rates, and greater learning outcomes.

The chapter begins with a discussion of e-learning vs. traditional education issues and a presentation of the e-learning tools and technologies currently available. Following this, the characteristics and types of e-learning computer-mediated communication (CMC) and online communities are explained, and the key players and student roles in e-learning communities are identified. The chapter continues with social interaction research in e-learning, concentrating on such important areas as factors that influence social interaction, peer support, student-centered learning, collaboration, and the effect of interaction on learning.

4.2　E-Learning

E-Learning can be defined as "learning that is supported by information and communications technologies" (Ward, 2003). Characteristics of e-learning include a physical distance between the students and teachers, and electronic technologies are usually used for the delivery of the material. The difference between distance learning and e-learning is the method of delivery. Distance learning can occur by mailing lecture materials to students in different countries, whereas e-learning is delivered using electronic means like the World Wide Web (WWW) although the terms are often used interchangeably. Traditional education, however, requires that the students and teachers be together in the same classroom where face-to-face teaching sessions are carried out.

E-Learning has been gaining public interest very quickly, as universities and businesses see it as an opportunity for cost savings and higher productivity. A study by the International Data Corporation (IDC, 2004) predicted that the e-learning market would grow at a compound average growth rate of 27%

annually from 2004 to 2008. The e-learning market was about $6.5 billion in 2003, and it is forecasted to increase to more than $21 billion by the end of 2008 (IDC, 2004). The Stanford Research Institute Consulting Business Intelligence Group (SRIC-BIG) suggests that this growth of the e-learning marketplace is a direct result of expanding the technology platform installed base (personal computers [PCs], compact disk-read only memory [CD-ROM], and Internet) and the enhanced capabilities provided by that platform over time (SRIC-BIG, 2002). SRIC sees the growth of the e-learning market as driven by wireless connectivity, simulation tools, learning object design, and peer-to-peer collaboration.

4.2.1 E-Learning vs. Traditional Education

There is no doubt that distance education is an exciting innovation. The idea of being able to work at your own pace from home sounds very appealing. Students may enroll at universities from around the world and take courses that their local universities do not offer without all the expense and hassle of traveling overseas. But is e-learning going to completely replace traditional education?

Souder (1993) argues that achievement on various tests administered by course instructors tends to be higher for distant as opposed to traditional students, whereas Egan et al. (1991) points out that conventional instruction is perceived to be better organized and more clearly presented than distance education.

As for the near future, it does not look like e-learning will completely replace traditional education. Videoconferencing helps to make lectures more like a classroom, where all the students can interact with each other and the tutor. However, videoconferencing and the other e-learning techniques cannot replace actual face-to-face contact. E-Learning should be seen as a complement to traditional education rather than a replacement.

DLBOIS (2002) states that by using distance learning as an add-on to traditional classroom education or as a replacement for some courses, institutions can develop a richer overall learning environment available to a greater population base.

4.2.2 E-Learning Technologies and Tools

Various technologies and tools are used to deliver online education. Each of these has different uses, along with advantages and disadvantages. Which tool to pick usually depends on needs, budget, and what equipment the students will have and be able to use. The following are the main types of technologies and tools used in e-learning (DLBOIS, 2002):

4.2.2.1 Instructional Television

Instructional television is a delivery system whereby programs are delivered using television broadcasting. It can be a one-way process where the programs have been prepared in advance, or a two-way process in which there is live interaction between the instructor and the students (Oliver, 1994). The pros of instructional television are that most people are already familiar with the medium and that video and audio can be represented. The cons are that the programs cannot be easily edited and that video production can be quite costly and time consuming. An example of an instructional television program is a biology lesson that teaches students about the feeding habits of a specific animal in its natural habitat.

4.2.2.2 Computers

Computers are very useful tools in distance education. They have the ability to integrate print, audio, video, and interactivity (Maier et al., 1998). Using computers has many advantages. They are interactive and can increase national or local access by the use of networks. Another advantage is that the users can set their own learning pace when using computers for learning. Constantly advancing and changing technologies promise many innovations but on the other hand can also be a disadvantage when trying to stay current. Other limitations include high software and network costs.

4.2.2.3 The Internet

The Internet is currently a widely used medium for distance education (Taylor, 2002). It automatically integrates all the previously mentioned computer features into a 24/7 globally accessed medium. E-mail, bulletin boards, chatting, dialogues, newsgroups, research, and interactive conferencing are all easily available with the Internet. Many courses are now offered online through a Web site. The WWW is very important because it enables the teacher to include content such as course information, assignments, tests, and lecture notes, and allows communication with the students by either e-mail or live conferencing. Online materials also include journals, articles, databases, software libraries, past examination papers, frequently asked questions (FAQs), and notice boards.

4.2.2.4 Interactive Videoconferencing

Interactive videoconferencing uses multimedia elements, digital cameras, and microphones to capture video and sound and transmit it live in real time to other users, who will receive it using their display units and speakers (Maier et al., 1998). When both sides have the necessary equipment, two-way communication can occur, making the videoconferencing an interactive process (Oliver, 1994). Wider uses of this technology can have multiple users and facilitate meetings or classrooms where the teacher and students are all connected from their homes. The main benefit of interactive videoconferencing is that being at the same location is no longer a restraint for having real-time visual communication. Other benefits include access to people in different parts of the world and the ability to include multimedia aspects during the conferencing. The biggest drawback of interactive videoconferencing is the high prices for the equipment and connections. At the moment, not many people have both the correct equipment and connection speeds to be able to carry out interactive videoconferencing.

4.3 E-Learning Communities

One of the most important characteristics of the Internet is the opportunity it offers for human-human communication through computers and networks. As Metcalfe (1992) points out, communication is the Internet's most important asset and e-mail is the most influential aspect. E-mail is just one of the many modes of communication that can occur through the use of computers. Jones (1995) points out that through communication services like the Internet, Usenet, and bulletin board services that are electronically distributed and almost instantaneous, electronic communication has for many people supplanted postal service, telephone, and fax machine. All these applications in which the computer is used to mediate communication are called computer-mediated communication (CMC).

4.3.1 E-Learning CMC and Online Communities

Metz (1994) states that CMC has been in existence since 1969 and provides a general definition of CMC as "any communication patterns mediated through the computer." (pp. 32). December (1997, p. 1) defines CMC as "a process of human communication via computers, involving people, situated in particular contexts, engaging in processes to shape media for a variety of purposes."

CMC is a broad area, and it is more accurately defined by its specific applications. In my study, I am focusing on Internet-based e-learning CMC. December (2004) has provided a definition of CMC that is oriented to this particular context:

"Computer-Mediated Communication (CMC) is the process by which people create, exchange, and perceive information using networked telecommunications systems (or non-networked computers) that facilitate encoding, transmitting, and decoding messages. Studies of CMC can view this process from a variety of interdisciplinary theoretical perspectives by focusing on some combination of people, technology, processes, or effects. Some of these perspectives include the social, cognitive/psychological, linguistic, cultural, technical, or political aspects; and/or draw on fields such as human communication, rhetoric and composition, media studies, human-computer interaction, journalism, telecommunications, computer science, technical communication or information studies" (December, 2004, p. 1).

Examples of e-learning CMC include asynchronous communication such as e-mail and bulletin boards; synchronous communication such as chatting; and information manipulation, retrieval, and storage through computers and electronic databases (Ferris, 1997).

E-Learning CMC has its benefits as well as its limitations. For instance, a benefit is that the discussions are potentially richer than in face-to-face classrooms (Scotcit, 2003), but on the other hand, users with poor writing skills may be at a disadvantage when using text-based CMC (Scotcit, 2003). Furthermore, asynchronous discussions allow for "reflective study followed by complex exchanges and genuine collaboration in the application of theory" (Sumner and Dewar, 2002, p. 1).

When it comes to Web site designers, choosing which e-learning CMC to employ (for instance, forum or chat room) is not a matter of luck or randomness. The determining factor when selecting a CMC is whether the communication is synchronous or asynchronous. In the case of e-learning, the choice of the appropriate mode of CMC will be made by answering questions such as the following (Bates, 1995; CAP, University of Warwick, 2004; Heeren, 1996; Resier and Gagne, 1983):

Are the users spread across time zones?
Can all participants meet at the same time?
Do the users have access to the necessary equipment?
What is the role of CMC in the course?
Are the users good readers/writers?
Are the activities time independent?
How much control are the students allowed?

Through the use of CMC applications, online communities emerge. As Korzeny pointed out early as 1978, the new social communities that are built from CMC are formed around interests and not physical proximity (Korzeny, 1978). CMC gives people around the world the opportunity to communicate with others who share their interests, as unpopular as these interests may be, which does not happen in the "real" world where the smaller the interest in a particular scene is, the less likely it will exist. This is due mainly to the Internet's connectivity and the plethora of information available posted by anyone anywhere in the world.

The term "online community" is multidisciplinary in nature, means different things to different people, and is slippery to define (Preece, 2000). The relevance of certain attributes in the descriptions of online communities, like the need to respect the feelings and property of others, is debated (Preece, 2000). Online communities are also referred to as cybersocieties, cybercommunities, Web groups, virtual communities, Web communities, virtual social networks, and e-communities, among other things.

For the purpose of a general understanding of what virtual communities are, Rheingold's (1993) definition is presented: "Virtual communities are social aggregations that emerge from the Net when enough people carry on those public discussions long enough, with sufficient human feeling, to form Webs of personal relationships in cyberspace" (p. 5). Rheingold further expands this definition by defining the net as "an informal term for the loosely interconnected computer networks that use CMC technology to link people around the world into public discussions" and cyberspace as "the conceptual space where words, human relationships, data, wealth, and power are manifested by people using CMC technology" (p. 5).

Due to the Internet and its related technologies, the population of these cyber-communities is global. The emergence of the so-called global village was predicted years ago as a result of television and satellite technologies by McLuhan (1964), although it is argued by Fortner (1993) that "global metropolis" is a more representative term (Choi and Danowski, 2002). It is estimated that as of September 2002 there are 605.60 million people online (Nuasoft Web Services, 2004). If one takes into account that the estimated world population of 2002 was 6.2 billion (U.S. Census Bureau, 2004), then the online population is nearly 10% of the world population. In most online communities, time, distance, and availability are no longer disseminating factors. Given that the same individual may be part of several different online communities, it is obvious why more and more online communities keep emerging and increasing in size. There are many things that bring people together in online groups. These include hobbies, ethnicity, education, beliefs, and just about any other topic or area of interest. Wallace (1999) points out that meeting in online communities eliminates prejudging based on someone's appearance, and thus people with similar attitudes and ideas are attracted to each other.

Cyberspace is the new frontier in social relationships, and people are using the Internet to make friends, colleagues, and lovers, as well as enemies (Suler, 2004). This statement pretty much sums up the importance of a person being part of an online community as well as the problems that may arise.

4.3.2 Key Players in E-Learning Communities

Preece et al. (2002) state that an online community consists of people, a shared purpose, policies, and computer systems while identifying the following member roles:

Moderators and mediators: guide discussions and serve as arbiters
Professional commentators: give opinions and guide discussions
Provocateurs: provoke other users
General participants: contribute to discussions
Lurkers: silently observe

There are four key players in e-learning communities:

Instructor: provides the content while considering the diversity of the audience (DLBOIS, 2002). The instructor should have a good knowledge and understanding of the technologies and delivery methods and should provide feedback to the students.
Facilitator: links the instructor with the students. The facilitator must understand the students and the instructor's expectations and "follow the

directive established by the teacher" (Distance education: An overview, 2002.)

Support staff and administrators: handle tasks such as student registration and grading, distribution of the material and copyright issues. Administrators make sure that the technology is used efficiently to meet the students' and academic institutions' needs (DLBOIS, 2002).

Students: learn (Behnke, 2003). Because the students will not have face-to-face interactions with the instructors and other students, they must learn to use the technologies that try to bridge this gap. Students can be broken down further into three student role groups (Laghos and Zaphiris, 2007; O Murchu, 2005):

Self-learner: These students need to see their own goals, organize their own work, and manage their own time (O Murchu, 2005). These students make connections with a large number of their peers. Their discussions are mainly on social topics and not course related. They prefer to learn on their own and mainly use the discussion boards to make friends and socialize with their peers (Laghos and Zaphiris, 2007).

Team member: These students work collaboratively. Their social interaction is in teams, and they are actively involved in their projects (O Murchu, 2005). They interact with a smaller number of their peers but more often, thus working more in small teams. The majority of their usage of discussion boards is for discussions about course material and helping out their peers (Laghos and Zaphiris, 2007).

Knowledge manager: The focus of the knowledge manager is on the development of knowledge products. These can be in the form of reports and newspapers. Their activities include searching for information, collecting and analyzing data, and designing reports (O Murchu, 2005). These students usually provide their own lecture notes, are connected with a large number of their peer students who depend on them for this material, and are part of a several cliques. Their contributions in the discussion boards are mainly on course-related material (Laghos and Zaphiris, 2007).

4.4 Social Interaction Research in E-Learning

In this section, I discuss research on social interaction in e-learning with the aim of gaining knowledge in what influences social interaction and how issues like collaboration and peer support can benefit learners.

4.4.1 Factors That Influence Social Interaction

There have been several studies that investigated the factors that influence social interaction in online courses. Vrasidas and McIsaac (1999) examined a university graduate online course in the use of telecommunications for instruction. The course consisted of eight students and one professor and was supported with FirstClass and a Web site with course material and other resources. Their data collection included observations that were tape-recorded, interviews, student work, and the teacher's mailbox messages.

Their analysis showed that four major factors influenced interaction: course structure, class size, feedback, and prior CMC experience (Vrasidas and McIsaac, 1999). For course structure, the authors mention that required activities and collaboration on peer editing of students' papers led to more interactions. Regarding class size, the authors concluded that there would have been more interactions in the course if there were more students enrolled in it. Furthermore, the students felt that the feedback they got in the course was not adequate, and this kept interaction levels low. Finally, the authors provide some insights into the relevance of prior CMC experience with interactivity. They found that students with no previous experience felt more comfortable using asynchronous communication, which gave them time to reflect on their ideas, as opposed to synchronous communication, which they found hard to keep up with. Students with prior CMC experience enjoyed both modes of communication and used emoticons more frequently.

The authors state that online interaction is solely constructed through language and suggest that educators should structure online courses for dialogue and interaction, that timely feedback should be provided to the students, and that students should be trained early in the course to use the conferencing systems and emoticons (Vrasidas and McIsaac, 1999).

In comparison to face-to-face interaction, "online interaction may be slower and 'lacking' in continuity, richness and immediacy ... however in some ways online interaction may be as good or even superior to face-to-face interaction" (Vrasidas and Zembylas, 2003, p. 1). Planned group activities can increase the feeling of social presence and learner-learner interaction (Vrasidas and McIsaac, 2000), and gaining access and status in a setting and socializing are examples of intentions that drive interaction (Vrasidas, 2002).

4.4.2 Peer Support and Student-Centered Learning

A type of interaction in online courses is peer support. "Peer support is a system of giving and receiving help founded on key principles of respect, shared responsibility, and mutual agreement of what is helpful" (Mead et al., 2001, p. 140). When people find others that they feel are like them, they feel a connection and a deep understanding based on mutual experience (Mead et al., 2001).

"Student centered learning is supported theoretically by various overlapping pedagogical concepts such as self-directed learning (Candy, 1991), student-centered instruction or learning (Felder and Brent, 2001), active learning (Ramsden, 1992), vicarious learning (Lee and McKendree, 1999) and cooperative learning

(Felder and Brent, 2001)" (Kurhila et al., 2004, p. 1). Examples of contemporary ways to support collaborative learning include awareness of others, joint building of knowledge, and matching unknown actors or resources; and these can lead to positive interdependence within the learning community as well as engagement, autonomy, and independence. It is important to have tools that allow easy and straightforward ways for community members to interact with and support each other in a peer-to-peer fashion (Kurhila et al., 2004).

Online peer support occurs using CMC. The importance of students learning from their study peers is increasingly being recognized by the e-learning community. "In some instances eLearning can foster a greater degree of communication and closeness among students and tutors than face-to-face learning" (Sumner and Dewar, 2002, p. 1).

Furthermore, studies show that students would prefer to contact their peer students (rather than their tutor) when they have difficulty with coursework, difficulty understanding lectures, and difficulty assessing facilities (Lockley et al., 2004). Thus, the peer support is an important aspect of e-learning in both the findings of researchers and the opinions of students.

4.4.3 Collaboration and Cooperation

In a study of the modes of learning, Chi et al. (1989) found that collaboration produces a 70% retention rate. Online courses are cited as having an average completion rate between 25% and 70%, and it was found that a key driver for completion is codependency (Chi et al., 1989).

It is suggested that learners be informed of who is present in the online sessions and how the group is composed in order to recognize each other and develop a sense of direction (Hamburg et al., 2003). The authors also emphasize the importance of e-moderators helping the students by initiating and supporting chats and online socializing, as this makes the students feel at home and more willing to contribute. Finally, they state that compared to individual and competitive learning, collaborative learning raises the students' achievement level and problem-solving activities and enhances the development of personal traits (Hamburg et al., 2003). Interaction benefits and motivates the learners and facilitates higher-order learning (McLoughlin, 2004).

Some authors argue that e-learning systems do not sufficiently acknowledge the importance of the social process and rely on passive material, limiting interactivity. They suggest that e-learning should be socially situated, thus providing active interaction with the users (Angehrn et al., 2001). Communities of practice (Wenger, 1998) is a theory of social learning in which learning is a social process rooted in specific social context. Although there have been numerous tools developed for supporting student collaboration, the use of facilities that sustain collaborative work by students in different locations are not emphasized in their environments (Fakas et al., 2005).

Fakas et al. (2005) proposed an electronic laboratory journal (eJournal) paradigm as a collaborative and cooperative environment for Web-based experimentation in education. The eJournal provides students with the Web-based tools needed to complete their assignments by discussing, exchanging, sharing, and documenting information (Fakas et al., 2005). These data chunks or fragments can be objects in any format, such as, text, graphics, and experimental results. The eJournal was evaluated by facilitating a group of engineering students at the School of Engineering of the Ecole Polytechnique Fédérale de Lausanne. The authors report that the fragments created by the students contained discussions, knowledge exchange, and sharing of outcomes (Fakas et al., 2005). Furthermore, they found the students' collaboration and participation encouraging because significant amounts of fragments were created by the students (Fakas et al., 2005).

4.4.4 Interaction and Learning

Studies show that interaction is a fundamental process for learning (Vygotsky, 1978; Vrasidas and McIsaac, 2000; Dewey, 1938) and that knowledge is constructed in communities of practice through social interaction (De Angeli and Sue, 2005). Social problems affecting online communities include social loafing, which leads to low participation rates, disinhibited behavior (like flaming and abuse), and diffusion of responsibility. A solution to this is the presence of a moderator who can reduce the antisocial behaviors that are triggered by anonymity (De Angeli and Sue, 2005). In addition, the authors believe that the sense of community is greatest for the student when there is a sense of connectedness with the course and it is engendered by both social and learning dimensions (De Angeli and Sue, 2005). They also note that people who interact more in an online course tend to achieve higher marks at exams, as opposed to lurking, which is not as successful (De Angeli and Sue, 2005).

Furthermore, learners perceive the content of communication as an information source (Aviv, 2000). The social interdependence theory of cooperative learning (Johnson and Johnson, 1999) suggests that the way social interdependence is structured determines how individuals interact, which in turn determines their learning outcomes. Cooperative experiences promote greater social support than do competitive or individualistic efforts (Johnson and Johnson, 1999), and stronger effects exist for peer support than for superior (teacher) support (Aviv, 2000).

4.5 Methods for Analyzing E-Learning Communities

In this section, I present the literature review on current methods for analyzing e-learning CMC activity. This was done in order to find out what the existing methods are, identify their characteristics, benefits, and limitations, and investigate whether or not they are applicable to e-learning environments and especially for the evolution of social networks in e-learning environments. As mentioned earlier, the Internet plays a vital role in socially connecting people worldwide (Laghos and Zaphiris, 2005a). The virtual communities that emerge have complex structures, social dynamics, and patterns of interaction that must be better understood. Using e-learning CMC, we are provided with a richness of information and pools of valuable data ready to be analyzed (Laghos and Zaphiris, 2005b).

Various aspects and attributes of e-learning CMC can be studied. For instance, the formation of social networks and their characteristics, including density, centrality, cores, and cliques (Laghos and Zaphiris, 2005). An approach to this is called social network analysis ([SNA] Laghos and Zaphiris, 2006; Wellman, 1997; Scott, 2002; Wasserman and Faust, 1994). Three important and widely used types of e-learning CMC analyses are content analysis, human-human interaction analysis, and human-computer interaction (HCI): analysis and are further explained in the next sections.

4.5.1 Content Analysis

Content analysis is a social science methodology in which recorded human communications are studied (Babbie, 2004). It is a technique for compressing many words of text into fewer content categories (Stemler, 2001; Weber, 1990). There have been several frameworks created for studying the content of messages exchanged in e-learning CMC. Examples include work from Archer et al. (2001) and McCreary's (1990) behavioral model, which identifies different roles and uses these roles as the units of analysis. Furthermore, in Gunawardena et al.'s (1997) model for examining the social construction of knowledge in computer conferencing, five phases of interaction analysis are identified: (I) sharing/comparing information; (II) the discovery and exploration of dissonance or inconsistency among ideas, concepts, or statements; (III) negotiation of meaning and coconstruction of knowledge; (IV) testing and modification of proposed synthesis or coconstruction; and (V) agreement statement(s) and applications of newly constructed meaning. Henri (1992) has also developed a content analysis model for cognitive skills, which is used to analyze the process of learning within the student's messages. Furthermore, Mason's work (1991) provides descriptive methodologies using both quantitative and qualitative analysis. In the case of e-learning, for example, a useful framework is the transcript analysis tool (Fahy, 2003), which offers the following advantages:

■ The approach is student centered.
■ It works with Gunawardena's model.
■ It was built on the weaknesses of other models.
■ It uses the sentence as the unit of analysis.

4.5.2 Human-Human Interaction Analysis

Over the years, researchers have used several techniques for analyzing interaction. It is important to note that the type of interaction studied in this case is interpersonal interaction, more specifically, the human-human interaction that takes place using e-learning CMC. Examples of interaction analysis models include Bale's interaction process analysis (Bales, 1950; Bales and Strodbeck, 1951), the Social Identity model of Deindividuation Effects (SIDE) model (Spears and Lea, 1992), a four-part model of cyberinteractivity (McMillan, 2002), Vrasidas's (2001) framework for studying human-human interaction in computer-mediated online Environments, and a technique called SNA.

4.5.3 HCI Analysis

A working definition of HCI as provided by ACM SIGCHI (2002, p. 8) is "Human-computer interaction is a discipline concerned with the design, evaluation and implementation of interactive computing systems for human use and with the study of major phenomena surrounding them." The focus is on the interaction between one or more humans and one or more computational machines (ACM SIGCHI, 2002). HCI is a multidisciplinary subject that draws on areas such as computer science, sociology, and cognitive psychology (Schneiderman, 1998). The concept of HCI consists of many tools and techniques that are used for information gathering and evaluation. Important HCI techniques include questionnaires, interviews, personas, and log analysis.

4.5.4 Evolution Analysis

Analyzing the evolution of e-learning communities is important because these people networks continuously evolve and change over time, and therefore keeping track of the network changes will enable educators to predict how certain actions will affect their network and to incorporate various methodologies to alter their state. One such evolution analysis method is FESNeL (Framework for assessing the Evolution of Social Networks in e-Learning.) FESNeL (Laghos, 2005) allows e-educators and online course instructors and maintainers to perform in-depth analyses of the communication patterns of the students of their e-learning courses and to follow their course's e-learning CMC progression. FESNeL assess the social network of the students over the duration of the course, thus mapping out the changes and evolution of these social structures over time. It is useful for monitoring the networks and keeping track of their changes while investigating how specific course amendments, participation in CMC, and/or conversation topics positively or negatively influence the dynamics of the online community. When using this framework to assess their e-learning community, e-educators are able to predict how certain actions will affect their network and can incorporate various methodologies to alter the state of their network.

FESNeL is a unified framework comprised of both qualitative and quantitative methods in which the unit of analysis is the social network. The four components of FESNeL are the Attitudes Toward Thinking and Learning Survey (ATTLS), SNA, topic relation analysis (TRA), and the Constructivist On-Line Learning Environment Survey (COLLES). An explanation of the FESNeL components follows.

4.5.4.1 SNA

"Social Network Analysis (SNA) is the mapping and measuring of relationships and flows between people, groups, organizations, computers or other information/knowledge processing entities. The nodes in the network are the people and groups while the links show relationships or flows between the nodes. SNA provides both a visual and a mathematical analysis of human relationships" (Krebs, 2004, p.1). Preece (2000) adds that it provides a philosophy and set of techniques for understanding how people and groups relate to each other. It is concerned about attributes between pairs of actors (such as kinship, roles, and actions) and has been used extensively by sociologists (Wellman, 1982, 1992), communication researchers (Rice, 1994; Rice et al., 1990), and others. Analysts use SNA to determine if a network is tightly bounded, diversified, or constricted, to find its density and clustering, and to study how the behavior of network members is affected by their positions and connections (Garton et al., 1997; Wellman, 1997; Scott, 2002; Knoke and Kuklinski, 1982).

4.5.4.2 TRA

The TRA model (Laghos, 2005; Laghos and Zaphiris, 2006) is a content analysis tool. Content analysis is a technique used in qualitative analysis to study written material by breaking it into meaningful units (Babbie, 2004). The data are collected directly from the discussion boards of an e-learning class and then sorted into the TRA categories. The TRA is a newly developed tool in which the units of analysis are the threads and messages of each of the discussions of the forum. The data collected include the messages per thread, the participants per thread, the discussion topic, and its relevance to the course. The tool assists us in understanding the messages and communication between the learners and how important the discussed topics are for the learners to remain and complete the online course. The TRA was developed to group the messages that the students post into categories, enabling us to determine which type of messages (peer support, off-topic conversations, etc.) engage the student most in the course and contribute to course retention. TRA is composed of three main categories, some of which have subcategories. These categories were deduced by observations of e-learning discussion boards and the different types of conversations that take place. The TRA categories are

A - Course material related
  Category A deals with conversations in the discussion boards of the e-learning courses that are related to the course material and is broken down into two further categories, A1 and A2.
A1 - Related to current lesson

Threads that belong in A1 are conversations that have to do with the course material of the current lesson. Examples of such topics include questions and answers and correcting peers' mistakes.

A2 - Related to course (but not current lesson)

Threads that belong in A2 are conversations that have to do with the course, but their subject is not in the current lesson's syllabus. For example, a conversation about an exercise in Lesson 3 posted in the discussion forum of Lesson 1 would go in this category. Also, a general question about mathematics (in an area that is not included in the mathematics lesson's syllabus) would go in A2.

B - Course Web site/technical related

Category B is specific to conversations regarding the course Web site and technical issues. Problems listening to audio files, accessing specific parts of the site, or usage issues are all in this category.

C - Not related to course

Finally, posts that are categorized in C are those that have nothing to do with the course or its usage. Category C has two subcategories.

C1 - Peer socializing

C1 is a broad category that covers conversation types in which peers socialize with each other. Examples include students introducing themselves, discussions about football games and concerts, and making new friends.

C2 - Other

Category C2 includes all the other off-topic conversations that are not about peers socializing with each other. Examples of posts that belong in this category are spam and advertisements.

4.5.4.3 COLLES

The COLLES measures students' perceptions and preferences and was designed to help teachers assess, from a social constructivist perspective, the quality of their online learning environment (Taylor and Maor, 2000). The COLLES electronic questionnaire was designed to support the use of the Web for teaching programs for which social constructionism is a key pedagogical referent and can be used to monitor the quality of innovative online teaching and learning (Taylor and Maor, 2000). It consists of twenty-four questions arranged into six scales (Dougiamas and Taylor, 2003):

Relevance: How relevant is online learning to students' professional practices?
Reflection: Does online learning stimulate students' critical reflective thinking?
Interactivity: To what extent do students engage online in rich educative dialogue?
Tutor support: How well do tutors enable students to participate in online learning?
Peer support: Do fellow students provide sensitive and encouraging support?
Interpretation: Do students and tutors understand each other's communications?

4.5.4.4 ATTLS

ATTLS is used to measure the quality of discourse within a course. It measures the extent to which a person is a "connected knower" (CK) or a "separate knower" (SK). People with higher CK scores tend to find learning more enjoyable and are often more cooperative, congenial, and willing to build on the ideas of others, whereas those with higher SK scores tend to take a more critical and argumentative stance to learning (Galotti et al., 1999).

The two different types of procedural knowledge (separate and connected knowing) were identified by Belenky et al. (1986). Separate knowing involves objective, analytical, detached evaluation of an argument or piece of work and takes on an adversarial tone that involves argument, debate, or critical thinking (Galotti et al., 1999). "Separate knowers attempt to 'rigorously exclude' their own feelings and beliefs when evaluating a proposal or idea" (Belenky et al., 1986, p.111). Separate knowers look for what is wrong with other people's ideas, whereas connected knowers look for why other people's ideas make sense or how they might be right, because they try to look at things from the other person's point of view and try to understand it rather than evaluate it (Clinchy, 1989; Galotti et al., 1999). These two learning modes are not mutually exclusive and may "coexist within the same individual" (Clinchy, 1996, p.207).

Differences in SK and CK scores "produce different behaviors during an actual episode of learning and do result in different descriptions of, and reactions to, that session" (Galotti et al., 1999, p.435).

4.6 Discussion and Conclusions

The purpose of this chapter has been to give a thorough description of the key concepts of e-learning and e-learning communities. It is important to understand the characteristics and technologies surrounding these areas before attempting to analyze the way students use them. To begin with, an overview of e-learning along with its technologies and tools were presented. In addition, e-learning CMC and the formation of online communities were discussed. Following this was an investigation into e-learning communities in which key players and student roles were identified.

Then, research about social interaction and communication in e-learning was presented, including the factors that drive interaction and a number of other interaction case studies. Research has shown that e-learning CMC is important for knowledge building. Studies have shown that interaction is influenced by four major factors: course structure, class size, feedback, and prior CMC experience. In addition, it has been identified that students prefer to contact their peer students rather than their tutor when they have difficulties with the online lessons. Student collaboration is seen to produce high retention rates, and online interaction with fellow peers has been linked with greater learning outcomes. These findings stress the importance of students interacting with each other when taking part in online learning courses.

Finally, a discussion of the methods used for analyzing CMC activity in e-learning environments was presented, commenting on their advantages and drawbacks. A methodological framework for studying the evolution of social networks in e-learning communities was also introduced, explaining each of its four components (SNA, TRA, ATTLS, and COLLES).

E-Learning is the future of education, and e-learning CMC technologies such as chat rooms and discussion boards should be incorporated into these environments to enable the students to interact and communicate with each other, as this promotes their sense of presence in the e-learning communities and positively influences their learning outcomes.

References

Angehrn, A., Nabeth, T., and Roda C. (2001). Towards personalised, socially aware and active e-learning systems. Centre for Advanced Learning Technologies (CALT) White Paper.

Archer, W., Garrison, R.D., Anderson, T., and Rourke, L. (2001). A framework for analysing critical thinking in computer conferences. European Conference on Computer-Supported Collaborative Learning, Maastricht, The Netherlands. March 22–24, 2001.

Association for Computing Machinery Special Interest Group Computer-Human Interaction [ACM SIGCHI]. (2002). *Curricula for Human-Computer Interaction*. New York: The Association for Computing Machinery.

Aviv, R. (2000). Education performance of ALN via content analysis. *Journal of Asynchronous Learning Networks*, 4(2): 53–72.

Babbie, E. (2004). *The Practice of Social Research*, 10th ed. Belmont, CA: Thomson/Wadsworth Learning.

Bales, R.F. (1950). A set of categories for the analysis of small group interaction. *American Sociological Review*, 15: 257–263.

Bales, R.F., and Strodbeck, F.L. (1951). Phases in group problem-solving. *Journal of Abnormal and Social Psychology*, 46:485–495.

Bates, A.W. (1995). *Technology, Open Learning and Distance Education*. London: Routledge.

Behnke. W. (2003). *Online/eLearning—An Overview*. Vancouver, Canada: Vancouver Community College.

Belenky, M.F., Clinchy, B.M., Goldberger, N.R., and Tarule, J.M. (1986/1997). *Women's Ways of Knowing: The Development of Self, Voice, and Mind*, 2nd ed. New York: Basic Books.

Candy, P. (1991). *Self-Direction for Lifelong Learning*. San Francisco, CA: Jossey-Bass.

CAP, University of Warwick. (2004). E-Guide: Using computer mediated communication in learning and teaching. Retrieved November 8, 2004, from http://www2.warwick.ac.uk/services/cap/resources/eguides/cmc/cmclearning/.

Chi, M., Bassok, M., Lewis, M., Reimann, P., and Glaser, R. (1989). Self-explanations: How students study and use examples in learning to solve problems. *Cognitive Science*, 13:145–182.

Choi, J.H., and Danowski, J. (2002). Cultural communities on the net—Global village or global metropolis?: A network analysis of Usenet newsgroups. *Journal of Computer-Mediated Communication*, 7:3.

Clinchy, B.M. (1989). The development of thoughtfulness in college women: Integrating reason and care. *American Behavioral Scientist,* 32:647–657.

Clinchy, B.M. (1996). Connected and separate knowing: Toward a marriage of two minds. In N. Goldberger, J. Tarule, B. Clinchy, and M. Belenky (Eds.), *Knowledge, Difference, and Power: Essays Inspired by Women's Ways of Knowing.* New York: Basic Books. pp. 205–247.

De Angeli, A., and Sue, K. (2005). Learning conversations: A case study into e-learning communities. Proceedings of the Interact 2005 eLearning and Human-Computer Interaction Worskhop. Rome, Italy. September, 12–16, 2005.

December, J. (1997). Notes on defining of computer-mediated communication. *Computer-Mediated Communication Magazine,* 3:1.

December, J. (2004). What is computer-mediated communication…? Retrieved October 19, 2004, from http://www.december.com/john/study/cmc/what.html.

Dewey, J. (1938). *Experience and Education.* New York: Collier Macmillan publishers.

Distance education: An overview. Retrieved November 13, 2002, from http://www.uidaho. edu/evo/dist1.html.

Distance Learning Benefits Organizations, Individuals and Society [DLBOIS]. (2002). Retrieved November 4, 2002, from http://www.ciscoworldmagazine.com/monthly/ 2001/04/distance.shtml.

Dougiamas, M., and Taylor, P. (2003). *Moodle: Using Learning Communities to Create an Open Source Course Management System.* EDMEDIA 2003. Honolulu, Hawaii.

Egan, M., Sebastian, J., and Welch, M. (1991). Effective television teaching: perceptions of those who count most … distance learners. *Proceedings of the Rural Education Symposium.* Nashville, TN.

Fahy, P.J. (2003). Indicators of support in online interaction. *International Review of Research in Open and Distance Learning,* 4(1).

Fakas, G.J., Nguyen, A.V., and Gillet, D. (2005). The Electronic Journal: A Collaborative and Co-operative Learning Environment for Web-Based Experimentation. Computer Supported Cooperative Work (CSCW): *The Journal of Collaborative Computing,* 14(3):189–216.

Felder, R., and Brent, R. (2001). Effective strategies for cooperative learning. *Journal of Cooperation and Collaboration in College Teaching,* 10(2):69–75.

Ferris, P. (1997). What is CMC? An overview of scholarly definitions. *Computer-Mediated Communication Magazine,* 4(1): 1–11.

Fortner, R.S. (1993). *International Communication: History, Conflict, and Control of the Global Metropolis.* Belmont, CA: Wadsworth.

Galotti, K.M., Clinchy, B.M., Ainsworth, K., Lavin, B., and Mansfield, A.F. (1999). A new way of assessing ways of knowing: The Attitudes Towards Thinking and Learning Survey (ATTLS). *Sex Roles,* 40(9/10):745–766.

Garton, L., Haythornthwarte, C., and Wellman, B. (1997). Studying on-line social networks. In S. Jones (Ed.), *Doing Internet Research.* Thousand Oaks, CA: Sage, 75–105.

Gunawardena, C., Lowe, C., and Anderson, T. (1997). Analysis of a Global Online Debate and the Development of an Interaction Analysis Model for Examining Social Construction of Knowledge in Computer Conferencing. *Journal of Educational Computing Research,* 17(4):397–431.

Hamburg, I., Lindecke, C., and Thij, H. (2003). Social aspects of e-learning and blending learning methods. Fourth European Conference E-COMM-LINE, Bucharest, Romania. September 25–26, 2003.

Heeren, E. (1996). Technology support for collaborative distance learning. PhD diss., University of Twente, Enschede, The Netherlands.

Henri, F. (1992). Computer conferencing and content analysis. In A.R. Kaye (Ed.), *Collaborative Learning through Computer Conferencing: The Najaden Papers,* 117–136. Berlin, Germany: Springer-Verlag.

International Data Corporation [IDC]. (2004). Growth of e-learning Market. IDC. Framingham, MA.

Johnson, D.W., and Johnson, R.T. (1999). *Learning Together and Alone, Cooperative, Competitive and Individualistic Learning.* Needham Heights, MA: Allyn and Bacon.

Jones, S. (1995). Computer-mediated communication and community: Introduction. *Computer-Mediated Communication Magazine,* 2(3):38.

Knoke, D., and Kuklinski, J.H. (1982). *Network Analysis.* Beverly Hills, CA: Sage.

Korzeny, F. (1978). A theory of electronic propinquity: Mediated communication in organizations. *Communication Research,* 5:3–23.

Krebs, V. (2004). An introduction to social network analysis. Retrieved November 9, 2004, from http://www.orgnet.com/sna.html.

Kurhila, J., Miettinen, M., Nokelainen, P., and Tirri, H. (2004).The role of the learning platform in student-centered e-learning. Proceedings of the Fourth IEEE International Conference on Advanced Learning Technologies, Joensuu, Finland. August 2005.

Laghos, A. (2005). FESNeL: A methodological framework for assessing the evolutionary structure of social networks in e-learning. Junior Researchers of the European Association for Research on Learning and Instruction (Eleventh Biennial JURE/ EARLI Conference), University of Cyprus, Nicosia. August 23–27, 2005.

Laghos, A., and Zaphiris, P. (2007). Investigating student roles in online student-centered learning. International Council for Educational Media Conference (ICEM 2007), Intercollege, Nicosia, Cyprus. September 20–22, 2007.

Laghos, A., and Zaphiris, P. (2006). Sociology of student-centred e-learning communities: a network analysis. e-Society 2006 Conference, Dublin, Ireland. July 13–16, 2006.

Laghos, A., and Zaphiris, P. (2005a). Computer assisted/aided language learning. In C. Howard, J.V. Boettcher, L. Justice, K. Schenk, P. Rogers, and G.A. Berg (Eds.), *Encyclopedia of Distance Learning* (Vol. 1, pp. 331–336). Hershey, PA: Idea Group Reference.

Laghos, A., and Zaphiris, P. (2005b). Online social structures and perceived attitudes towards thinking and learning. Fifth International Conference on the Scholarship of Teaching and Learning (SoTL), London, UK. May 12–13, 2005.

Laghos, A., and Zaphiris, P. (2005c). Frameworks for analyzing computer-mediated communication in e-learning. Eleventh International Conference on Human-Computer Interaction (HCI-International), Las Vegas, NV, June 22–27, 2005.

Lee, J., and.McKendree, J. (1999). Learning vicariously in a distributed environment. *Journal of Active Learning,* 10:4–10.

Lockley, E., Pritchard, C., and Foster, E. (2004). Professional evaluation: Students supporting students—lessons learnt from an environmental health peer support scheme. *Journal of Environmental Health Research,* 3(2): 74–81.

Maier, P., Barnett, L., Warren, A., and Brunner, D. (1998). *Using Technology in Teaching and Learning.* London, UK, Kogan Page.

Mason, R. (1991). Analyzing computer conferencing interactions. *Computers in Adult Education and Training,* 2(3):161–173.

McCreary, E. (1990). Three Behavioural Models for Computer-Mediated Communication. In L. Harasim (Ed.), *Online Education: Perspectives on a New Environment.* New York: Praeger. pp. 133–184.

McLoughlin, C. (2004). A Learning Conversation: Dynamics, Collaboration and Learning in Computer Mediated Communication. Perth, Australia. WA: TAFE Media Network.

McLuhan, M. (1964). *Understanding Media: The Extension of Man.* New York: McGraw Hill.

McMillan, S.J. (2002). A four-part model of cyber-interactivity: Some cyber-places are more interactive than others. *New Media and Society,* 4(2):271–291.

Mead, S., Hilton, D., and Curtis, L. (2001). Peer support: a theoretical perspective. *Psychiatric Rehabilitation Journal,* 25:134–141.

Metcalfe, B. (1992). Internet fogies to reminisce and argue at Interop Conference. *Info World.* September 1992 (21), p. 45.

Metz, J.M. (1994). Computer-mediated communication: Literature review of a new context. *IPCT, Interpersonal Computing and Technology: An Electronic Journal for the 21st Century,* 2(2):31–49.

Nuasoft Web Services. (2004). Nua Internet Surveys. Retrieved October 20, 2004, from http://www.nua.ie/surveys/how_many_online/index.html.

Oliver, E.L. (1994). Video tools for distance education. In B. Willis (Ed.), *Distance Education: Strategies and Tools.* Englewood Cliffs, NJ: Educational Technology Publications, pp. 165–198.

O Murchu, D. (2005). New teacher and student roles in the technology-supported language classroom. *International Journal of Instructional Technology and Distance Learning.* 2(2): 3–18

Porter, L.R. (2004). *Developing an Online Curriculum: Technologies and Techniques.* Hershey, PA: Information Science Publishing.

Preece, J. (2000). *Online Communities: Designing Usability, Supporting Sociability.* Chichester, UK: John Wiley and Sons.

Preece, J., Rogers, Y., and Sharp, H. (2002). *Interaction Design: Beyond Human-Computer Interaction.* New York: John Wiley and Sons.

Ramsden, P. (1992). *Learning to Teach in Higher Education.* London, UK: Routledge.

Reiser R.A. and Gagne', R.M. (1983). *Selecting Media for Instruction.* Englewood Cliffs, NJ: Educational Technology Publications.

Rheingold, H. (1993). *The Virtual Community: Homesteading on the Electonic Frontier.* Reading, MA: Addison-Wesley.

Rice, R. (1994). Network analysis and computer mediated communication systems. In S.W.J. Galaskiewkz (Ed.), *Advances in Social Network Analysis.* Newbury Park, CA: Sage, pp. 167–203.

Rice, R.E., Grant, A.E., Schmitz, J., and Torobin, J. (1990). Individual and network influences on the adoption and perceived outcomes of electronic messaging. *Social Networks,* 12:17–55.

Schneiderman, B. (1998). *Designing the User Interface: Strategies for Effective Human-Computer Interaction* (3rd Ed). Menlo Park CA: Addison Wesley.

Scotcit. (2003). Enabling large-scale institutional implementation of communications and information technology (ELICIT). Using computer mediated conferencing. Retrieved November 2, 2004, from http://www.elicit.scotcit.ac.uk/modules/cmc1/welcome.htm.

Scott, J. (2002). *Social Network Analysis: A Handbook.* 2nd ed. London, UK: Sage.

Souder, W. (1993). The effectiveness of traditional vs. satellite delivery in three management of technology master's degree program. *The American Journal of Distance Education.*

Spears, R., and Lea, M. (1992). Social influence and the influence of the social in computer-mediated communication. In M. Lea (Ed.), *Contexts of Computer Mediated Communication*, 30–65. London, UK: Harvester Wheatsheaf.

Stanford Research Institute Consulting Business Intelligence Group [SRIC-BIG]. (2002). *Technology Evolution in e-learning*. Stanford Research Institute Consulting Business Intelligence Group. Menlo Park, CA.

Stemler, S. (2001). An overview of content analysis. *Practical Assessment, Research and Evaluation*, 7(17). Available online at http://pareenline.net/getvn.asp? 6=7 and n=17.

Suler, J. (2004). The final showdown between in-person and cyberspace relationships. Retrieved November 3, 2004, from http://www1.rider.edu/~suler/psycyber/showdown.html.

Sumner, J., and Dewar, K. (2002). Peer-to-peer elearning and the team effect on course completion. *ICCE*: 369–370.

Taylor, R. (2002). *An Introduction to e-learning*. Creative Learning Media. Uxbridge, UK.

Taylor, P., and Maor, D. (2000). Assessing the efficacy of online teaching with the Constructivist On-Line Learning Environment Survey. In A. Herrmann and M.M. Kulski (Eds.), *Flexible Futures in Tertiary Teaching. Proceedings of the Ninth Annual Teaching Learning Forum, 2–4 February 2000*. Perth, Australia: Curtin University of Technology.

U.S. Census Bureau (2004). Global population profile 2002. Retrieved October 20, 2004, from http://www.census.gov/ipc/www/wp02.html.

Vrasidas, C., and McIsaac, S.M. (1999). Factors influencing interaction in an online course. *The American Journal of Distance Education*, 13(3):22–36.

Vrasidas, C., and Zembylas, M. (2003). The nature of technology-mediated interaction in globalized distance education. *International Journal of Training and Development*, 7(4):1–16.

Vrasidas, C., and McIsaac, M. (2000). Principles of pedagogy and evaluation of Web-based learning. *Educational Media International*, 37(2):105–111.

Vrasidas, C. (2002). A working typology of intentions driving face-to-face and online interaction in a graduate teacher education course. *Journal of Technology and Teacher Education*, 10(2):273–296.

Vrasidas, C. (2001). Studying human-human interaction in computer-mediated online environments. In Y. Manolopoulos and S. Evripidou (Eds.), *Proceedings of the Eighth Panhellenic Conference on Informatics*, Nicosia, Cyprus, November 8–10, 2001, 118–127.

Vygotsky, L.S. (1978). *Mind and Society: The development of Higher Mental Processes*. Cambridge, MA: Harvard University Press.

Wellman, B. (1997). An electronic group is virtually a social network. In S. Kiesler (Ed.), *Culture of the Internet*, 179–205. Hillside, NJ: Lawrence Erlbaum.

Wenger, E. (1998). *Communities of Practice; Learning, Meaning and Identity*. Cambridge University Press.

Wasserman, S., and Faust, K. (1994). *Social Network Analysis: Methods and Applications*. Cambridge University Press.

Weber, R.P. (1990). *Basic Content Analysis*, 2nd ed. Newbury Park, CA.

Ward, C. (2003). E-Learning training: Catching up with the future. EDUCAUSE 2003 Conference. May 6–9, 2003, Adelaide, Australia.

Wallace, P. (1999). *The Psychology of the Internet*. Cambridge, UK: Cambridge University Press.

Wellman, B. (1982). Studying personal communities. In P.M.N. Lin (Ed.), *Social Structure and Network Analysis*. Beverly Hills, CA: Sage. pp. 61–80.

Wellman, B. (1992). Which types of ties and networks give what kinds of social support? *Advances in Group Processes*, 9:207–235.

Chapter 5

Technology-Enhanced Health Communities
Research on the Use of Social Computing Applications by Both the Sick and the Well

Laura Slaughter, Aksel Tjora, and
Anne-Grete Sandaunet

Contents

5.1 Introduction

Health-related social computing is often discussed in light of one major change in the health care industry. This key change we are referring to concerns the role of patients, which has shifted from "passive recipient of care" to "active partner" in maintaining health and managing illness (Cahill 1998). People have become consumers of health care services, and their responsibility for making decisions and collaborating with health care professionals has increased. Many people have embraced this role, whereas others are left feeling unqualified to make decisions or unable to cope (Hoffman 2005). However, this expectation of consumerism in health care is here to stay, and it influences and is influenced by people's use of the Internet as a source of information and support.

This leads us to one concept that is increasingly seen in the health informatics literature, "patient empowerment." Demiris (2006) writes, "Empowerment can be perceived as an enabling process through which individuals or groups take control over their lives and managing disease." It involves self-awareness, personal responsibility, informed choices, and quality of life (Feste and Anderson 1995, p. 186). Patient empowerment is generally seen as positive, as are the changes that have resulted in viewing people as consumers of health care. Health care researchers and professionals are actively trying to find ways to facilitate patient involvement in care. The Web, and especially social computing, is seen as a vehicle for empowering patients and helping them become "active partners" (Donnelly et al. 2008; Polomano et al. 2007). Many, such as Pratt et al. (2006) suggest that patients need tools that address current information gathering and sharing challenges, thereby enabling them to be more involved in their health care. Bos et al. (2008) believe that "the self-care information tool of the future will be a combination between the patient's observation record and the Internet, with the doctor and the patient positioned together at the intersection but not having to pay attention to the technology."

It is thought that Web 2.0, which refers to technologies that encourage social interactivity of Web-delivered content and services, will help the health care industry "move into the 21st century" (Deshpande and Jadad 2006, p. 1). The terms "Health 2.0" and "Medicine 2.0" have been coined in reference to social computing in health care, use of Web 2.0 technologies, and the changes taking place in modern health care systems. Hughes et al. (2008) have defined "Medicine 2.0" as "the use of a specific set of Web tools (e.g., blogs, podcasts, tagging, and wikis) by actors in health care including doctors, patients, and scientists, using principles of open source and generation of content by users, and the power of networks in order to personalize health care, collaborate, and promote health education." Eysenbach (2008) proposes a triangle of participants: the patient, the health care provider, and the biomedical research scientist, each having a stake in the information exchange related to health. His triangle is visionary. In the past, health consumers and professionals were segregated and primarily communicated within their own separate groups, but now this division is disappearing. As Briceño et al. (2008, p. 1037) emphasize, there is a connection now between researchers, patients, and clinicians: "Information that was once the private property of a privileged few can now be available to all. Reciprocal and multi-directional communication is now possible between organizations, and between researchers, clinicians, policy makers, patients, and the general public."

The focus of this chapter is on the patient/caregiver users of social computing in health. They are referred to in the health care informatics literature as "health consumers," and they are a diverse group, ranging from people who have an acute problem (e.g., broken leg) or a health-related concern (e.g., what kind of diet will help them lose weight) to those who have a long-standing chronic illness and the terminally ill. They are usually laypersons in the domain of health and medicine, although some may have acquired near expert level knowledge of the specific disease they are diagnosed with or problem they are experiencing. Why this focus? There is awareness that individuals as well as society as a whole has much to gain from health consumers' use of social computing applications. The goals are to involve people in health in order to improve people's overall health, conduct research more efficiently, lower costs and wastefulness, and increase the quality of services.

This chapter will describe the types of technologies currently used for health-related social computing. We then discuss the characteristics of users and their motivations for forming virtual health communities and interacting socially online. We present both the positive and the problematic issues that arise. Social computing can be highly beneficial, helping people to cope in difficult situations, but it also presents certain drawbacks, such as the possibility of getting bad advice from nonexperts on tough medically complex matters. We then review some of the research conducted within the health sciences, viewing social computing technology as a medical intervention that can affect health outcomes (e.g., social contacts helping cancer patients feel less depressed and anxious.) We then shift focus to health care professionals' use of social computing tools and show how health consumers are becoming involved in clinical research through these technologies. The final sections look at (1) medical sociology, an important research field explaining the effects of health-related social computing; and (2) ethical issues that arise when conducting social computing research involving health consumers.

5.1.1 Brief Overview of Health Social Computing Technologies

As we look across the range of literature describing technology use for supporting health-related communication and virtual community formation, we find much relevant work occurring well before the Web 2.0 era. Initially, email mailing lists (listservs), message or bulletin boards (asynchronous discussion forums), and chat rooms (synchronous discussion forums) were the three main ways that people in virtual communities communicated with one another using the Internet. These facilitated interaction, exchange, and community building. They are still the focus of social computing research studies currently being published. For example, in 2008, Stoddard et al. (2008) published a study looking at the effects of adding a bulletin board–based virtual community service within a Web-assisted tobacco intervention site.

The development of more recent technologies that have been associated with Web 2.0, such as blogs, wikis, social networking sites, and three-dimensional (3D) virtual worlds, have resulted in a wave of research on how they are used by the "worried well," patients and caregivers, as well as health care workers. Most of these Web technologies are generic in the sense that they can be used by anyone for any purpose, not just health topics. Health care professionals and researchers have discussed their potential for use related to health, believing that these tools promote collective participation beyond what the Web previously had to offer. For instance, social networking sites such as Facebook™ (http://www.facebook.com) can provide emotional support in times of stress through "weak ties" to others, i.e., relations that are easily maintained with little input (Granovetter 1973) and lacking the intimacy and frequency of interaction characteristic of family and close friends (Bender et al. 2008).

Within health care, recent efforts have been made to design and develop tools specifically for patients and caregivers. One example of this is the Web site "CarePages" (http://www.carepages.com). CarePages are bounded blogs, having a login that connects patients with friends, family, and other concerned individuals. They provide a means to disseminate information such as status and future treatment plan in an efficient way. Other tools such as "patient portals" and "personal health records" (PHR) are in the forefront of health informatics research. Patient portals are usually institution-sponsored Web sites that allow patients and caregivers to communicate with health personnel directly, pay bills, make appointments, and access parts of their health records (e.g., test results.) There are many different notions of how PHRs should function and what features they should support, but these are essentially tools that allow people to store and maintain their own electronic health record. They are usually described as applications that support sharing health information in the PHR with health personnel and others the patient has authorized (Tang et al. 2006). Both patient portals and PHRs tend to have a built-in service for patient-to-clinician interaction, which is very popular with users. In a study of one patient portal called PatientSite, the email clinical messaging feature was one of the most frequently used tools within the portal (Weingart et al. 2006).

Table 5.1 provides an overview and brief description of the technologies available for social computing relevant to consumer health users. We also give an example of a research publication related to each technology type in order to illustrate the diversity of work in this area.

Table 5.1 Common Types of Social Computing Applications Used by Health Consumers

Technology	Description	Research Example
Mailing lists	A form of asynchronous communication where email messages sent by individuals are distributed to a list of subscribers.	Rimer et al. (2005) studied how new subscribers use cancer-related online mailing lists. Their data suggest that mailing lists are perhaps underused by minority patients.
Discussion groups, bulletin boards	A type of asynchronous messaging, nowadays often viewed from a Web browser. Users can post messages by emailing to the group, and others can respond. The discussion is public.	Use of discussion groups in health care has been researched extensively. Bar-Lev (2008) detailed analysis of "emotion talk" among members of an HIV/AIDS support group.
Instant messaging	A type of synchronous communication used between two or more people. Usually text is typed and conveyed through devices connected over the Internet.	Researchers have tested the use of instant messaging and presence technology for the location of experts and the management of their time affecting the deployment of eHealth services in the WebOnCOLL system (Chronaki et al. 1997).
Chat rooms	A form of synchronous conferencing allowing users to share an online space. Chat rooms can be either text only or multimedia (3D virtual space).	Woodruff et al. (2001) use a virtual world chat room as a part of a smoking cessation program in which young smokers who live in rural areas can interact (i.e., chat) in real time with a trained smoking cessation facilitator as well as with each other.

(continued)

Table 5.1 (Continued)   Common Types of Social Computing Applications Used by Health Consumers

Technology	Description	Research Example
Blogs	A blog is a Web site that contains entries, like a diary, and is usually kept by one individual. People reading blog entries are able to leave comments in response to the entry. Blogs can contain elements that are typical of any Web site, including images and text.	Blogs created by health consumers often chronicle a person's experience with a specific disease. Health care professionals have used blogs to learn more about patient experiences. Mehta (2007) reported on sources of the anxiety of patients with uveitis in order to improve physician counseling of these patients.
Wikis	A wiki is a Web site that allows users to directly modify content and add content using a simple mark-up language.	Daub et al. (2008) report on an RNA annotation wiki for the molecular biology research community.
Social networking sites	These are Web-based services that allow users to interact with each other through the creation of "profiles," instant messaging, and email messaging. Users create an online community within a password-controlled set of Web pages.	Dong et al. (2008) studied the romantic communication between users of MySpace™, a social networking service. They looked at the relationship between romantic communication and self-esteem, self-image, and emotional intelligence.
Social bookmarking	These are services for storing references to Web sites and files available through the Internet. Users tag the references, or bookmarks, in order to organize them for searching and browsing. The collection of tags (or keywords) created by users is often called a "folksonomy."	Smith and Wicks (2008) studied the tags added by patients to describe their symptoms on an online social networking site. They compared the keyword tags used by patients to the terms available in a health care professional thesaurus.

(continued)

Table 5.1 (Continued) Common Types of Social Computing Applications Used by Health Consumers

Technology	Description	Research Example
Web-based shareable and distributed electronic health/patient records	These technologies include PHRs, PHAs, and patient portals. These are Web-based services that let people store their own health data, track their health (or manage a specific condition), and make contact with other patients/caregivers.	Frost and Massagli (2008) discuss a PHA Web service called "PatientsLikeMe," which is an online community built to support information exchange between patients. They identified and analyzed how users of this platform reference personal health information within patient-to-patient dialogues.
3D virtual world	These are 3D multimedia environments where users can create an avatar and manipulate elements of the virtual world. They can include both text and voice communication. These worlds might also be considered 3D chat rooms, with their main function to provide a simulated environment for interactions.	Gorini et al. (2008) discuss the role played by 3D virtual worlds in psychotherapy applications. Their paper discusses "p-health": personalized immersive e-therapy whose key factor is interreality.

Note: HIV/AIDS, human immunodeficiency virus/acquired immunodeficiency syndrome; PHA, personal health applications; PHR, personal health record.

5.2 Characteristics of Health Social Computing Users

We first discuss some general characteristics of users of health-related social computing, concentrating on the health consumer, which includes the "patient," "caregiver," "citizen with health concern," or "lay health user." People who make use of social computing are a subset of those who use the Internet for health purposes, and therefore, they share some of the same characteristics. Results of the Pew Internet and American Life surveys of Americans' Internet use show a rising percentage of Internet users seeking information for health purposes. In 2000, 46% of adults in the United States had access to the Internet and 25% had looked online for healthcare information. This increased to 74% online, 61% looking for health information in 2009. Users are "most likely women, those under sixty-five, college graduates, those with more online experience, and those with broadband access" (Fox and Jones, 2009). Forrester research published a report in 2008 on the adoption of social technologies by online health information seekers in the United States and "found them to be more engaged on average in social computing than U.S. online consumers in general. (http://www.forrester.com/Research/Document/Excerpt/072114607100. html)" Within Europe, women were also found to be the most active users of eHealth services. Factors that are associated with the use of eHealth services, from a study covering seven countries (Denmark, Germany, Greece, Latvia, Norway, Poland, and Portugal) were "youth, higher education, white-collar or no paid job, having a visit to the General Practitioner during the past year, long-term illness or disabilities, and a subjective assessment of one's own health as good" (Andreassen et al. 2007).

The fact that women use the Web as a source of health information more than men is not overly surprising, because in general, women are known to be the "gatekeepers" for health information (Warner and Procaccino 2004). As Stern (1986 p. xi) states, "Women have guarded the health of their families since the dawn of time." What we also see is that Web users, on average, are often the most highly educated and have the best overall health status. These demographics may change in the future as newer social computing tools become ubiquitous and as more people, not just those with current health concerns, begin to see value in the use of specific health management applications.

One of the key concerns of health care researchers is "vulnerable populations." These are people who are more susceptible to "falling through the cracks" in today's health care system, which requires patients to be active in acquiring information and resources. The elderly, disabled, and young children are often considered vulnerable populations, as are the poor, immigrants, those with mental disorders, and the homeless. One way to help vulnerable populations is to improve equality of Internet access so that they too can receive benefits from the health information available and an ability to communicate via Web-based tools. Some of the main obstacles to overcome are cultural differences, limited literacy skills, low education levels, and lack of access to technology (Cashen et al. 2004). Gilmour (2007) has discussed some methods for helping to alleviate inequalities in Internet access, which in turn, sustain health inequalities. These include providing free Internet services at strategic health sites; improving document readability and cultural acceptability of health information on Web sites; and teaching the computer skills required to navigate Web sites effectively.

5.2.1 Consumer Health Users of Specific Social Computing Applications

The majority of research that has been conducted on consumer health social computing has been focused on disease-specific support groups (i.e., using discussion systems, chat rooms, and mailing lists.) These are old technologies. Vast numbers of articles have been published about the formation of social networks on the most popular technology used by support groups, the Internet discussion system (e.g., Usenet newsgroups.) Most of these virtual communities form around a specific diagnosis, for example, colon cancer. Users are primarily those who experience symptoms and are seeking a diagnosis or have been diagnosed, or they could be the friends, relatives, or caregivers of an ill person. Other groups might form around specific types of health concerns that are not necessarily related to the user or a significant other being diagnosed with an illness, including immunization, pregnancy, weight loss, and smoking cessation.

People participate in health online support groups for a number of reasons. Some of the major goals of patient users are to decrease their sense of alienation, anxiety about treatment, and feelings of isolation (McKenna et al. 1995). Many people turn to support groups in order to clear up misconceptions and get information that will allow them to make decisions and achieve their health goals. Users' messages on these groups have been found to contain (1) questions (needs for information), (2) information given in response to specific questions, (3) personal opinions, (4) encouragement/support, (5) personal experiences, (6) thanks, (7) humor, and (8) prayer (Klemm et al. 1998). The frequency of these message types can differ by disease, stage of disease, and other factors. For instance, users who have been diagnosed with a chronic disease might request educational information more often than they discuss emotional stress. They might have come to the group seeking to learn from others' experiences. Caregivers of terminally ill patients, on the other hand, have strong needs for emotional support (Buis 2008). One research study specifically looked at the content of posts from people with a rare disease, primary biliary cirrhosis. They compared message frequency and content from different stages of the disease. What they found, contrary to their hypothesis, was that newly diagnosed people do not post more messages overall. "At different points in the illness, people with primary biliary cirrhosis have different concerns to communicate" (Lasker et al. 2005). Ginossar (2008) looked at gender differences in participation as well as the role of family members in support seeking and provision.

The above discussion covers studies of health consumer participation on Internet discussion groups. However, users can form virtual communities using a wide variety of available social computing technologies (see Table 5.1). Some of these systems are designed as a tool for any topic and others for health-specific use, as discussed above (section 1.1). Each has features that can make them more or less attractive to users as a vehicle for communication. As we have seen, there are specific needs being fulfilled through participation in health discussion groups. Moore and Serva (2007) looked at motivational factors for participating in virtual communities: wikis, blogs, and Internet forums. They propose a framework that posits specific motivational factors for each medium. For example, altruism (concern and benevolence), collaboration (assisted articulation of ideas), egoism (peer recognition), reputation (social standing and status), and self-esteem (respect and positive reinforcement) can all be motivational factors for wiki use. Their work was conducted in a business setting with the main goal of designing software packages that facilitate business-oriented virtual communities. The framework is untested in the domain of health virtual communities. This leaves opportunities for further research on understanding health-related motivations for using different types of social computing applications. They might have missed specific motivations that are important to people who have health concerns, for example, the ability to contribute anonymously can affect choice of tool used. Consider, for example, a woman who finds out she is pregnant and might not want her co-workers to know immediately. She might turn to using a chat room with her friends instead of posting the information on her social networking site profile.

When it comes to research about specific types of technology used by health consumers, more and more studies are being published about the use of blogs. Who are these users and what do they gain from this form of public journal writing? Bloggers can be people who simply want to tell their own "health story" about personal experiences. They might also be "consumer health reporters" who function as journalists on health care issues. For the latter group, many types of patient support Web sites featuring blogging functions have been set up by hospitals; the most commonly known is the "CarePages.com" service, which provides personalized Web sites "where members relate their stories, post photos and update friends and family instantly; in turn, people who care send messages of love and encouragement" (McDaniel and Duckler CarePages FAQ, p.1). Journalistic "reporters" use public blogs, and are more interested in generating discussion on biomedical topics. Guadagno et al. (2008) hypothesized that specific personality traits are associated with people who blog. They found that people who are high in openness to new experience and high in neuroticism are likely to be bloggers. However, their research was conducted on bloggers in general, not specifically on health blogging. On the surface, it seems that CarePages bloggers may have different personality traits from "journalistic" bloggers, although this hypothesis has not been tested.

Adams (in press) notes, "it is crucial to examine from a user's perspective what aspects of these blogging tools and alternative site options are most important and why/how they are effective. This should include first identifying and categorizing health-specific blogging sites and typifying who uses different types of blogging sites, as well as why and how they use them." So far, we have very few of these types of studies available to answer all our questions about bloggers. Kovic et al. (2008) published a study on the "medical blogosphere" and examined blog writers who behave essentially as journalists of medical issues, noting "only a small fraction of our bloggers constructed their blogs around their personal lives," and they were not concerned with sharing information with family or friends. Most of these bloggers were males between the ages of thirty and forty-nine, and a high percentage held a master's degree or doctorate. They spent a great deal of effort to write their blogs, checked sources, and worked to ensure quality information. Their major motivations for doing so were "sharing practical knowledge and skills as well as influencing the way other people think." This was one of the few studies that directly explored blog users. However, some blog user characteristics are available due to the research studies of clinicians who have used them as data sources for improving health services. Mehta (2007) examined patient blogs for descriptions of anxiety in uveitis patients (an ocular inflammatory disease.) They found 103 patient blogs. Only nine bloggers revealed their ages, which were between eighteen and seventy. Of the seventy-three bloggers who revealed their gender, the majority of them were women (forty-six out of seventy-three.)

Similar to blogs, the number of publications concerning the use of other recently developed social computing technologies, specifically portals and PHRs, is rapidly increasing. Although, one of the possible barriers to publishing about these users is patient privacy, many hospitals have been sponsoring portal Web sites in the past few years; but they might not publish their internal statistics on site usage. What makes a person interested in using a patient portal, for example? What features should social computing sites for health consumers have? Zickmund et al. (2008) provide some useful results. What they found was that patient dissatisfaction with health care provider communication and responsiveness, inability to obtain medical information, and problems communicating with office staff was linked to their desire to use a patient portal. Using the portal, patients got around their problems of receiving timely notice of test results and getting an appointment. Yet, when patients already had a good relationship with their health care provider (in this case, their general practitioner), they were not motivated to use the portal. Weingart et al. (2006) published a study on users of PatientSite, an Internet portal that allowed patients to ask nonurgent questions about care or symptoms using a Web messaging service, review radiology and pathology test results, request an appointment, renew prescriptions, obtain medical information, and view or add comments to their medical records. What they found was that those patients having the fewest medical problems, the young, and the affluent were the most frequent users of the site. Although, they state that the "enrollees were not entirely homogeneous, 7% of the users were at least age 65 (p. 94)."

We briefly presented results of two patient portal studies, whose functionality was similar. There was an emphasis on communication between the patient and the health care provider(s). It is possible that other types of social computing tools with additional functionality will be highly valued by certain populations. For example, Zimmerman et al. (2008) introduce a collaborative medical history portal that allows families to capture and store medical history information that is relevant to breast cancer. With these available data, the system can automatically generate risk assessments for breast cancer. The PatientsLikeMe (http://www.patientslikeme .com) site contains tools for patients to exchange their own health data (e.g., symptoms experienced) with each other. Frost and Massagli (2008) report that patients were generally asking for advice from another user with a particular experience, offering advice to a user with a specific symptom or health problem, and fostering relationships based on shared attributes. We are only beginning to see the results of health consumers using more sophisticated social computing applications such as PatientsLikeMe.

5.2.2 The Case of an Online Self-Help Group for Women with Breast Cancer

There is one major difference between general users of social computing and those who use these tools for health reasons. That is, health consumer site users are often in situations that provoke a great deal of anxiety and are life disruptive. Imagine, for instance, new parents who learn that their newborn has a heart defect and are desperate for emotional support and information. Logging on to a social networking site for health-related matters is not a hobby for many health consumer site users; it is imposed on them by misfortune. Even in the case of the "worried well" and the "curious well" (e.g., athletes interested in nutrition), there are very serious health implications involved for these users that can influence their quality of life. Below, to illustrate the impact social networking has on patients' well-being, we present the case of breast cancer patients who made use of a discussion service (Sandaunet 2008).

To study the effects of discussion group interaction on breast cancer patients, a new support group Web site was established and accessible through the Web site of the Norwegian Cancer Association between October 2003 and March 2005. Forty women enrolled in the study, and during a period of fifteen months a total of 1,114 messages were posted, with an average of 2.5 messages each day the group was accessible. Data were obtained through a questionnaire, by means of participant observation of the activity in the group, and through qualitative interviews. This study was privileged in the sense that not many researchers who do Web-based research have the opportunity to interview patients who post to a discussion list.

This online self-help discussion group was actively used and highly appreciated by a group of study participants. These were mainly women who were under treatment for the disease, particularly women whose cancer had spread. What changes or effects resulted with the discussion forum? First, group participation contributed to the adoption of a consumer identity in the communication with health care professionals, by giving the women a greater sense of control. Second, group participation provided hope for a qualitatively rich life even as cancer was spreading, helping the women to redefine this situation as a situation with meaning. This was helpful both for those who were diagnosed with spreading and for those who feared that they later could find themselves in this situation. Women in the group who had experienced spreading of the cancer became role models for others. Third, group participation can break the sense of isolation that was particularly experienced by those with spreading. For these women, it is reasonable to argue that alternative ways of being ill were facilitated through group participation. On the whole, then, the study can be argued to give some resonance to proposals of these types of Web sites as a unique and global space in which people can rewrite or reconstruct their illness narrative (Hardey 2002). It is particularly important to note how it provided a space for acknowledging the insecure future.

On the whole, having the discussion group facilitated access to other women with spreading. Sharing experiences and knowledge appeared as particularly crucial for the helpful function of this group. As Walther (2004) points out, the Net is about not only *what* you get but also *who* you get. The social networks provided by Web use may differ radically from those of offline life. This one had a number of women participants who have experienced the spreading of breast cancer. The helpful function of the group toward women who had experienced spreading is worth noting. More people now live longer with cancer (Cancer in Norway 2006). In this material, those who were diagnosed with spreading also participated in "normal" activities, such as entering into a role as an active mother or grandmother. Some were even able to continue their professional work, at least for short periods. As such, people who are living with their incurable condition are becoming more "visible."

The anonymity of the Internet was not particularly emphasized among these women, whereas the written communication was commented on as advantageous because of its therapeutic benefits (Pennebaker and Beall 1986). It is further interesting how the flexibility of the Internet was commented on by some of these women as an increased opportunity to talk about the illness when the need was apparent and not at a particular point of time, which is the case in face-to-face–based groups.

5.2.3 Lay Users in a Complex Expert Domain

Emotional support and changes in patient empowerment are often the central themes in research concerning users of social computing related to health. However, it is not just emotional support that motivates people to interact online. Trying to understand medical treatments, making decisions, and managing care are all topics that come up for users of health-related social computing. Searching for information means using technology to come into contact with "peer experts" or at least people who are currently going through the same experiences. One of the key problems for people with health concerns is that they are laypersons, whereas the topic area that they deal with can be complex. A full knowledge of bodily functions, effects of treatment, medication interactions, and other information requires years of medical training.

The majority of discussions have focused on the possibility that participants who are not medially trained will spread incorrect information within support groups, on blogs, and on wikis (Culver et al. 1997; Adams in press). This is often corroborated by reports of adverse events due to Web site misinformation and the poor quality of health-related Internet resources in general (e.g., Black and Hussain 2000). There are a number of articles that address the dangers of incorrect information on the Internet, with a focus on Web sites, not necessarily online groups and social computing tools (Eysenbach and Köhler 2002; Bessell et al. 2002). These studies have not found supporting evidence that this is a widespread problem (Crocco et al. 2002), but results are mixed concerning the medical correctness of health information generated by laypersons. Esquivel et al. (2006) examined 4,600 posts on a breast cancer support group, noting the errors and the rate of correction. Only ten postings (.22%) were found to be false or misleading, and seven of these were quickly identified and corrected by other participants within an average of 4 hours and 33 minutes. Tsai et al. (2007) collected a set of postings related to diabetes mellitus type 1 from three discussion sites. They found that 48% of postings contained medical content, and 54% of these were either incomplete or contained errors. Van Uden-Kraan et al. (2008) analyzed the contents of 1,500 messages from eight Dutch support groups (three breast cancer, three arthritis, and two fibromyalgia.) They did not find a single post that contained medical information that was "potentially dangerous to others." The majority of the messages having medical content (11% of all messages) discussed conventional treatments, and most of the sources of the medical information written in posts came from personal experiences and personal communication.

Other related work in this area is centered on patient language, expressions used to describe illness, and ways of communicating health care concerns. Some of this work falls within psychology (Weinman and Petrie 1997), health communication, and, for example, narrative medicine (Pownall 2004). Researchers with a technology development interest have also looked closely at language use characteristics in order to create ways to automatically process texts produced by health consumers (Leroy 2006).

All the work on health consumer language has implications for developers who wish to integrate, for example, patient social computing with health professionals' clinical systems. Zeng et al.'s (2002, p. 289) report on the differences between laypeople and professionals, states that "mismatches are not only lexical in nature, but also semantic, and that consumers employ a different mental model from clinicians with respect to medical terminology" and further that "linking of complex medical concepts with a consumer's mental model is a new challenge for health IR." Smith and Wicks (2008) studied folksonomic tags for medical terms added by users on the Web site PatientsLikeMe. These users either have or know someone with amyotrophic lateral sclerosis, multiple sclerosis, or Parkinson's disease. Forty-three percent of PatientsLikeMe symptom terms were found to be exact matches (24%) or synonymous (19%) to terms in professional medical vocabularies from a well-known source, the Unified Medical Language System (UMLS; http://www.umlsks.nlm.nih.gov). Slightly more than half did not match the UMLS or were unclassifiable.

5.3 Social Computing as a Medical Intervention

Within health sciences research, the use of social computing technology can be viewed as a type of intervention. That is, it can have some type of a health effect on the user in the same way that a new drug or some other type of treatment might have. The assumption is that social interaction can help patients to cope better with being ill through both emotional and information support. This, in turn, should lead to improved health outcomes, for example, alleviation of depressive symptoms. In the past, Internet discussion groups were used to test this assumption. Eysenbach et al. (2004) reviewed forty-five publications looking specifically at the effect of virtual communities on health outcomes. They conclude, "despite extensive searches in the health, social sciences, communication, and informatics literature we failed to find robust evidence on the health benefits of virtual communities and peer to peer online support. (p. 1169)" According to Eysenbach, there are several problems with these studies that may have prevented researchers from finding evidence of an effect. First, the studies most often combine complex interventions, for example, some educational component along with the discussion group, making it difficult to draw conclusions based on the effectiveness of "pure" electronic peer-to-peer interactions. Second, the study participants are not representative of real-life users of discussion groups and as Eysenbach points out, "participants may need to have the intrinsic desire to communicate with other people in order for virtual communities to be beneficial. (p.1169)" Third, and this is probably related to the second problem, there was often a lack of participation in the virtual community components of these studies, making it difficult to show an effect if there is one.

Since Eysenbach reported on these studies, studies of virtual communities and their effects on health outcome are changing—because the technology is changing. Newer tools can allow data exchange between patients as well as integration with professional researchers' databases. In future, we hope to see studies that focus on patients' use of social computing to mobilize and coordinate with researchers in clinical studies. One of the more recently reported incidents focused on a community using the social networking site PatientsLikeMe. PatientsLikeMe gives patients the ability to exchange information, including customized disease-specific outcome and visualization tools that help patients aggregate data on health outcomes. In the study reported by Frost et al. (2008), patients from the amyotrophic lateral sclerosis community initiated and participated in a kind of patient-driven trial of the use of lithium to stop disease progression. These patients used the site to monitor and report on the effectiveness of lithium during the trial.

5.3.1 Psychotherapeutic Interventions

One of the most natural ways to use social computing technology in health care is to help people who experience psychosocial challenges. A few studies report on the theme of blog writing as a form of self-therapy. Disclosing stressful or traumatic experiences through expressive writing is related to well-being (Smyth and Helm 2003). Blogs hold particular promise as a tool for those who write as a means to understand their problems and work through difficult or traumatic situations. Being ill with a chronic or serious disease can be one such trauma. The idea of writing as self-therapy is not new; in fact, it has been long known that diary writing can be beneficial. The one important difference between the use of blogs and private diary writing is that bloggers become part of a community and can receive responses to their entries. Tan (2008, p. 154) presents a picture of a blogger, Jenny, who found that "making the shift from the private world of pen and paper expressive writing to blogging was significant." She checks comments from readers on her blog regularly. Blogging is a medium that combines diary writing and support groups. Future research on psychotherapeutic benefits might examine the emotional effects of blogging on patients who are dealing with stressful illnesses where, for example, the treatment outcome is uncertain.

3D virtual worlds hold potential for helping people with psychosocial challenges to try out and practice new skills within a safe environment. The hope is that they will transfer their newly acquired social skills to the real world. *Second Life* (http://secondlife.com) is a popular place to conduct research studies. It is a 3D virtual world where members create their own avatar and can communicate through voice and text chat. One of the many examples described in Gorini et al. (2008) tells about the case of a medical student suffering from agoraphobia (the fear of crowds and traveling to places outside where one feels in control.) By creating an avatar that was close to his appearance, he could see himself navigating in threatening places and become comfortable within the virtual world before venturing out into the real. He successfully used *Second Life* to become more comfortable with unfamiliar and open spaces. Other 3D environments have been built for studies on specific patient populations. One example of this is Zora, an environment built at the Massachusetts Institute of Technology media laboratory to help pediatric patients on hemodialysis for end-stage renal disease to form a therapeutic virtual community, explore their identity, cope with having their illness, and provide mutual support (Bers et al. 2002).

5.3.2 The Dark Side

Even though there are plenty of positive aspects related to social computing by health consumers, there are also possible harmful health effects. Some researchers have expressed the possibility that people will be less likely to form or maintain meaningful real-life relationships because of their involvement online (Cummings et al. 2002). Another fear is based on prior studies on the lack of social constraints and feelings of depersonalization occurring in online groups (Sproull and Kiesler 1986). There have been assumptions that this would also occur within health- or illness-specific support groups and that those who seek support may experience negative or hostile retaliative posts (e.g., flaming; Kim and Raja 1991; Finfgeld 2000). In earlier writings concerning online health support groups, concerns were voiced about this "dark side of participation" (Preece and Mahoney-Krichmar 2003; Jadad 2006). Their two major concerns were that participants who were not medically trained would spread incorrect information within these groups and that people would self-diagnose rather than visit a health care professional. An even "darker" suggestion is that people with malicious intent can purposely mislead people, for example, in a hoax, causing harm to vulnerable patients/caregivers (Ebbinghouse 1998). Voiced concerns over privacy violations have also appeared in the literature (Eysenbach 2001).

A number of studies report that the dangers described above are minimal (Eysenbach 2001), although, we still find work that points to negative effects. Malik and Coulson (2008) report that discussion group participation can result in negative feelings resulting from news that others in the group have died (grief) or have left due to successful treatment (resentment and jealousy). Anderton and Valdiserri (2005) connected outbreaks of sexually transmitted diseases with men who meet other men in online chat rooms. There is still doubt whether participation in social computing is safe in relation to group processes where members can egg each other on to take specific actions. Whitlock et al. (2006) discuss how teenagers who self-injure through cutting themselves can normalize and encourage these behaviors on discussion group sites.

5.4 Professional Use of Health Social Computing

We briefly touch on the use of social computing by health care professionals in this section. We include both clinicians and researchers in this discussion. Social computing is of interest to health care professionals for connecting to other professionals as well as to health consumers (e.g., some forums have physicians on staff to answer patients' questions and monitor discussions.) Many forms of social computing provide a rich source of data that can be used to improve the quality of care, help inform clinical research, and warn of possible outbreaks (Jones and Alony 2008). Social computing technologies are increasingly used in medical education, including podcasting (Sandars et al. 2008; McKinney and Page 2009) and sites for career-advancing social networking among students (Tilley et al. 2006; Thompson et al. 2008). Use of 3D virtual worlds has been successfully tried as a method to teach cardiopulmonary resuscitation (CPR) to high school students (Youngblood et al. 2007). These worlds can also be a place where psychologists and sociologists can study human behavior.

An illustrative example of a clinicians' virtual community is "Sermo" (http://www.sermo.com), which is designed as a forum for physicians to keep them "on the cutting edge of medicine." They can post questions, share ideas, and learn from each other's experiences. Some tools are developed for specific communities, like the "Swedish Oral Medicine Network," who have been discussing patient cases since the mid-1990s through telephone conferences. This existing community recently developed SOMWeb "a semantic-web based system for supporting collaboration of distributed medical communities of practice." The clinicians' wish list of requirements for the system included features such as the ability to support knowledge reuse, data exchange, and reasoning based on ontologies (Falkman et al. 2008). It is obvious, based on SOMWeb, that clinicians are expecting more out of their tools than simply computer-mediated communication. They want to create and share content, and they need clinical decision support. In an interesting article, Wright et al. (2009) describe three case studies of Web 2.0 methods for developing decision support. One of these is the Clinfowiki (http://www.clinfowiki.org/). "The Clinfowiki is designed to be a resource for people interested in clinical decision support to both learn about as well as share their knowledge and content. The idea is that by involving as many disparate people, with as many different opinions as possible, extremely high-quality information and content will emerge. (p. 336)"

Sharpton and Jhaveri (2006) assert that "online communities hold the promise to enhance scientific research." Biomedical researchers have leveraged existing tools to benefit and increase collaboration and networking. Facebook™, commonly used for friends and family, has been used by professionals in the drug discovery community (Bailey and Zanders 2008). Generic researcher-oriented social bookmarking tools (not just for health care researchers) are often used, for example, Connotea (http://www.connotea.org/) has been popular in recent years. This tool is a reference management system that allows researchers to apply a descriptive identifier, a tag, which allows them to organize their references. Other users can then search these tags to find articles their peers have indexed as being important. Researchers often make use of wiki technology for project management and as a working environment. Daub et al. (2008) describe community annotation by researchers in molecular and cellular biology using a wiki. They added six hundred articles on a wiki devoted to noncoding RNA (ribonucleic acid, a molecule that consists of a long chain of nucleotide units), and the community was invited to update, edit, and correct these articles.

Newer technologies are being developed for researchers, for instance, the Scientific Collaboration Framework (SCF), which has been designed to leverage existing knowledge repositories and make use of knowledge and annotation available on the Web using an ontology (Das et al. 2009). SCF software has been used to create PD Online (http://www.pdonlineresearch.org/), a "large self-organizing community of basic and clinical scientists, industry professionals, and grant-makers, dedicated to advancing the treatment, prevention and cure of Parkinson's disease." The community knowledge base infrastructure will capture online discourse and integrate it with public biological databases for access that is freely available to all. Technologies such as this one go beyond what is considered "Web 2.0"; they use "Semantic Web" (Wikipedia contributors 2009) resources that define the semantics of information and make it possible to reason and do "smart things" with Web content.

To conclude this short section, we show one way that epidemiology and public health benefits from advances in social computing. "HealthMap" (http://www.healthmap.org) is an example of a Web site that gathers and displays information about infectious disease outbreaks across the globe. It works by taking Web-accessible information sources, such as news sources and disease reporting networks, and representing these reports on a map of the world. The user can quickly get an overview of places where an infectious disease has been reported. Checkboxes and sliders allow the user to filter and select certain news feeds or the specific diseases they want to see on the map. These data can be used and integrated with other geographical data using mapping software (Lefer et al. 2008). Besides mapping news feeds and content, future systems might use health consumer–reported data within social networking sites to alert public health officials to outbreaks. This, however, would come at the price of users' privacy because it would require the monitoring of content on these sites.

5.5 Medical Sociology and Health-Related Social Computing

The use of social computing technologies for health-related purposes has been of growing interest within the field of "medical sociology" or "sociology of health and illness." In particular, sociologists maintain a critical interest in whether, and potentially how, the Internet might contribute to such things as the democratization of health knowledge, the deprofessionalization of medicine, and patient empowerment (Broom and Tovey 2008). Both the health consumers' and professionals' use of social computing is sociologically relevant, not to mention systems for interaction between these two groups of users.

Relevant perspectives for the sociology of health-related social computing will be those we might term "social constructionism" and "interpretivism." Within these frameworks we do not take straightforward "functionalist" descriptions for granted, i.e., what works and what does not. For example, in a review of consumer health information seeking on the Internet, Cline and Haynes (2001) mention navigation difficulties (e.g., information overload, disorganization, lack of user-friendliness) and hazards (e.g., lack of peer review, misleading information) as important problems. A constructivist perspective will be more concerned with how users' interpretation of health-related use of the Internet may lead, for instance, to changed behaviors or changes in the understanding of the relation, meaning, and importance between lay and professional knowledge.

Within medical sociology, the question of whether increased access to medical domain knowledge for the health consumer results in a power gain or loss, in terms of their relationship to professionals, is of major importance. This debate deals with the hypothesis of "medicalization": that various concerns or deviances are increasingly regarded as medical problems that must be treated in medical institutions (Conrad 1992; Conrad and Schneider 1992; Illich 1976; Zola 1972). The medicalization process may be influenced by health-related use of the Internet, particularly virtual communities and social networks. Inspired by the French sociologist/philosopher Michel Foucault, Lupton (1998) suggests that a "medical gaze" is mediated through the Internet users' actions, hence giving biomedical terms and models greater impact in people's everyday lives. However, other sociologists, like Williams and Calnan (1996) suggest a "re-skilling" hypothesis in which laypeople's medical and health-related knowledge increases when they access and discuss health-related matters on the Internet. This, then, represents a form of "de-medicalization." The Internet, with its blogs and virtual communities, represents a quite different way for health consumers to share experiences that formerly was available only for those admitted to a hospital ward (Album 1996). As often happens within sociology, there are conflicting theories of the consequences of Internet technologies. Nevertheless, sociologists will agree that health-related use of the Internet can be regarded as a possible supplement to other (more professionally and politically regulated) health services but that there are no clear-cut "effects," but rather uncertain shifts in power, positions, knowledge hierarchies, responsibilities, expectations, and so on. It still leaves us with the questions of where, when, and what with regard to the resources consulted by people with health problems (both with and without the Internet)?

A sociologically informed analysis of the impact of social computing health services should be somewhat sober. People have always left only minor parts of their health troubles in the hands of the clinician and discussed most of them instead with family and friends, their communication medium for doing so being irrelevant. The anthropologist Arthur Kleinman studied how medicine and other cultures of society differentiate between which problems are medically relevant (Kleinman 1980). Based on comparative studies of Chinese and Western medicine in Taiwan in the 1970s, Kleinman defines three different (but overlapping) sectors within health care: the popular, the professional, and the folk. It is in the lay, nonprofessional, nonspecialist, popular sector in which illness is first defined and health care activities initiated. Kleinman (1980) suggests that between 70 and 90 % of all illness episodes are managed within the popular sector. Although people in Norway, for example, have good access to a doctor, about 80% of people's health problems are solved without involving their help (Grimsmo 1988). As Kleinman stated, the customary view is that professionals organize health care for laypeople. However, laypeople typically activate their health care by deciding when and whom to consult, whether or not to comply, when to switch between treatment alternatives, whether care is effective, and whether they are satisfied with its quality (Kleinman 1980). Hence, it is within the popular sector, among laypeople, that most decisions on health interventions are made, on the basis of knowledge about illness within the family, personal experiences, stories of illness and cure, beliefs about sickness, and so on.

This is where health-related social computing gets sociologically interesting, despite this aforementioned sociological soberness toward technological novelties. According to Kleinman (1980), health-related decisions are made within the layperson's concepts of illness rather than a biomedical understanding of disease.

However, as various sources of biomedical literature, knowledge, and research are made available on the Internet, biomedical concepts of disease may influence lay understanding of sickness to a larger extent. As noted by Conrad and Schneider (1992), when medical perspectives of problems and their solutions become dominant, they diminish competing definitions. However, interestingly enough, in a study of users of open access health forums on the Internet, Tangen and Tjora (2002) found that the impact of medical knowledge gained from Internet use was ambivalent—the findings pointing in various directions. In observations of interaction in discussion groups, it was shown that on the one hand, lay users seemed to point each other to biomedical Web pages, but on the other hand, physicians as moderators of discussion groups acknowledged lay users' advice.

In general, when there is a well-developed trust relation between lay and professional actors, Web-based and email-based communication seemed not to alter power relations very much (Andreassen et al. 2006; Tangen and Tjora 2002). Different expectations of involvement from lay and professional actors, however, seem to be a source of conflict. Andreassen and Trondsen (2008) have demonstrated that general practitioners who started using e-mail to communicate with patients were overwhelmed by the fact that patients did not want to use this opportunity only for simple messages (such as changing appointments or renewing prescriptions) but also for elaborating on illness stories and experiences. It has also been demonstrated (Broom 2005b, 2005c) that some medical specialists experience online cancer patient support groups as a threat to their expert status and control over decision-making processes. In the case of prostate cancer patients, Broom has shown there are differing experiences within online support groups. Such groups have provided some prostate cancer patients with a method of managing constraints posed by society's dominant constructions of masculinity, allowing for increased sharing and intimacy by limiting inhibitions associated with face-to-face encounters. However, other men view online support groups as havens for deception and misinformation and computer-mediated communication as a highly problematic form of social interaction (Broom 2005b).

In the constructivist sociology, it is argued that health services are produced in the interaction between patients and providers (Tjora 2008); both the providers' role as professionals and the patients' role as laypeople are key to the notion of "health service." In this situation, deprofessionalization denotes a trend toward a demystification of medical expertise and increasing lay skepticism about health professionals, suggesting a decline in the power and status of the medical profession. The emergence of Internet resources is one of the processes that are suggested to be relevant for this possible development, in addition to increasing consumerism and the rise of complementary medicine. According to Broom (2005a) the deprofessionalization thesis is inadequate for capturing the complex and varying ways in which specialists view and respond to the Internet-informed patient. Broom suggests that the specialists' adapting to the Internet and the Internet user should be viewed as strategic responses, rather than reflecting a breakdown in their authority or status. He even suggests that medical experts might be employing disciplinary strategies that reinforce traditional patient roles and alienate patients who use the Internet (Broom 2005c). In a study about cancer patients' use of the Internet to access information about complementary and alternative medicine (CAM), Broom and Tovey (2008) found a potential role of the Internet in reinforcing biomedicine's paradigmatic dominance in cancer care (i.e., medicalization).

The sociology of health-related social computing is therefore critically and theoretically ambivalent toward the democratic promises of the Internet. Although there is a certain democratic potential in providing media for interaction between patients to share experiences, the dominating role of physicians to regulate access to health services is a crucial factor in the effectiveness of these communication media.

5.6 Ethical Issues

As is evident in this chapter, researchers from a variety of disciplines conduct studies that involve collecting data from health-related social networking sites and virtual communities. For example, the design of these sites might be informed through the study of user characteristics, leading to improved tools and features. This would motivate collecting data regarding users' preferences for communication and data exchange. Within a 3D virtual world, a sociologist might be interested in observing social behaviors, or a psychologist might want to use a specific room or space to understand more about group therapy. Clinical researchers might analyze a collection of blogs or discussions produced by a specific patient group in order to find out more about drug side effect complaints. There are several possibilities arising for data collection. Existing texts or other multimedia data can be collected from a publically available site, for example, texts of discussions that will be automatically categorized. A researcher might become a member of an existing community to observe the interactions in that place, possibly participating themselves in the community. The researcher at that point might actively approach people in the community to gather face-to-face data; for instance, the researcher might want to interview participants in depth. Yet another option is that a community Web site is set up in which people are enrolled in a research study before they receive access to the site (i.e., only research participants are part of the virtual community.)

Although social networking sites and the Internet in general make it possible for researchers to gain access to social interactions and vast amounts of health consumer produced texts, the use of this resource raises new issues in research ethics. Above all, the researcher has an obligation to ensure that their work does not cause people harm. People who participate in health-related virtual communities are vulnerable; they are perhaps seriously ill, worried about their health, or worried about the health of a loved one. Some people post personal information out of desperation that they would otherwise keep private. Others may not realize that what they post will be stored and publically accessible for years to come.

There is a debate concerning whether or not written texts to groups and other traces of interactions between virtual community participants are "fair research game." That is, are they "public" or "private," and do they require informed consent to be analyzed as a part of a research study? One argument is that much of what is available through chat rooms, blogs, and other social computing tools are public discourses and are analogous to other public records, archives, letters in newspapers, etc., which can be freely used for research purposes. On the other side of this debate are those who believe that these sources cannot always be considered public because there are important psychological differences—participants never intended their postings to be completely open and did not realize the possible implications of participation. Eysenbach and Till (2001) documented negative reactions and resentment from participants in discussion groups when they realized a researcher had been monitoring their group, with some members leaving the group due to these feelings. Yet, it is also possible, especially in the case of medical research that has the potential to improve patients' quality of life or even find cures, that group members might welcome the presence of researchers. Some patients can actively seek out clinical researchers to participate, if they are pursuing new courses of treatment and do not wish to wait for researchers to "do it the hard way" by organizing full-fledged clinical trials (Frost 2008).

Another question regards the need for consent from "the enterprise" (e.g., Facebook is an enterprise) to analyze data available on their site. Many health institutions, such as hospitals or clinics, provide Web sites for patients to communicate with each other, family members, or their professional care team. These almost certainly require a login, making them "private" arenas for interaction. Even if the health care institution does not match log-ins to patients being treated, researchers should check with the institution concerning their research policy. Other sites without log-ins with discussion forums (e.g., a patient advocate institution with an open blog discussion) require a check of the enterprise's policy on reuse of contents for research purposes. Copyright is also an issue, both for enterprise content and, especially, for blogs that are in the style of "medical/health journalist."

Whether permission is deemed necessary for analyzing user contributions or not, publications reporting results must protect privacy. One concern in health social computing research is that very seriously ill people can be easy to identify due to their circumstances, especially if they have a unique case—and researchers must be sensitive to this. Sometimes simply knowing the gender of the contributor or region where they live can be enough to identify a patient with some very rare disorders. It is possible to find posts on support groups where people have described their surgical history and medications they are taking along with their email address and physical location. Researchers that publish direct quotes from these posts need to consider that even if they omit the "personal information" (e.g., email address), it is still possible for anyone reading the article to use a search engine to find the original message.

Another ethical issue involves researcher participation. Researchers can, for example, conduct naturalistic observations in public places without asking for consent. Is a 3D virtual world such a public space? What about the avatar researchers might create in order to conduct their observations? Is it all right for researchers to represent themselves as something they are not in order to conduct their study (e.g., create an avatar of an ethnic or racial background that is not their own)? Should researchers be allowed to create blogs representing themselves as cancer patients if they do not actually have cancer? These questions and the potential of the researcher to alter and possibly damage a functional health care–related virtual community must be considered within the framework of the study's purpose.

There are various forms of possible danger to participants who might be enrolled in social computing studies, and these are issues researchers must be aware of as they design their research protocols. Recruiting minors to research studies through a social networking site can be a challenge. Normally, research studies require parental consent, but the reality is that most teenagers surf the Internet unsupervised. Flicker et al. (2004) addressed the problem in the TeenNet project and discuss the processes that lead to allowing youth users to consent to participation without their parents' permission. The researchers set up a site that posed minimal risks to youths but that provided them with highly beneficial, prescreened, and reliable health information. In other research domains, particularly in psychotherapy research, there are potential dangers to participants. Some enrollees might become susceptible to the addiction of using the Internet, for instance. For some therapeutic situations, use of an "alternative world" could cause significant distress. As we can see, there are many ethical considerations in conducting health-related social computing studies, some of these questions being critical to protect people who may be more vulnerable than the average social computing user.

5.7 Summary

This chapter began with the concept of patient empowerment and the idea that health consumers, researchers, clinicians, and others (policy makers) can now engage in reciprocal and multidirectional communication through social computing technology. We have discussed what kinds of users form virtual communities and what their motivations are for doing so, from altruism (concern and benevolence) to egoism (peer recognition.) We have seen that health consumers can use social computing technologies to get personalized help, emotional support, and encouragement during difficult health situations. In addition, this chapter covered how social computing can affect health outcomes—through writing blogs as a form of self-therapy to social engagement as a possible way to reduce feelings of isolation in breast cancer patients whose cancer has spread. The "dark sides" of social computing were explored; we discussed the possibility of, for instance, people within a network encouraging risky or dangerous behaviors. We touched briefly on social computing by clinicians and researchers, with the notion that one day professional and health consumer technologies will become highly integrated. Lastly, we introduced sociology theory to work in the area, adding what is known about the effect of these technologies on, for one, the deprofessionalization of medicine.

In all, we have covered a great deal in a short amount of space. The technologies themselves were not a critical focus for the chapter—technology is changing rapidly and will continue to do so. We expect that future systems will allow greater exchange of data, both between health consumers and health care professionals. These tools will provide better ways for patients to support one another and allow them to connect to health care institution services when necessary. Social computing holds the potential for benefiting society at both the individual and group levels. By individual-level benefits, we mean the possibility of personalized medicine and better care for patients. At the group level, data gathered from social networking sites might be used to advance treatment for specific diseases. That is, large-scale clinical trials can take place to collect data from social computing applications and merge it with other clinical information. Future research in this area should focus on data integration and social computing between health consumers and health care professionals.

References

Adams, S. in press. Blog-based applications and health information: Two case studies that illustrate important questions for Consumer Health Informatics (CHI) research. *Int J Med Inform*, Corrected Proof.

Album, D. 1996. *Close Strangers: Patient Culture in Hospitals* [in Norwegian]. Oslo, Norway: Tano A.S.

Anderton, J. P., and R. O. Valdiserri. 2005. Combating syphilis and HIV among users of Internet chatrooms. *J Health Commun* 10 (7):665–71.

Andreassen, H. K., M. M. Bujnowska-Fedak, C. E. Chronaki, R. C. Dumitru, I. Pudule, S. Santana, H. Voss, and R. Wynn. 2007. European citizens' use of E-health services: A study of seven countries. *BMC Public Health* 7:53.

Andreassen, H. K., and M. Trondsen. 2008. Patient on E-mail. In *The Modern Patient* [in Norwegian], edited by A. H. Tjora. Oslo, Norway: Gyldendal Akademisk, p. 122–136.

Andreassen, H. K., M. Trondsen, P. E. Kummervold, D. Gammon, and P. Hjortdahl. 2006. Patients who use E-mediated communication with their doctor: New constructions of trust in the patient-doctor relationship. *Qual Health Res* 16 (2):238–48.

Bailey, D. S., and E. D. Zanders. 2008. Drug discovery in the era of Facebook—new tools for scientific networking. *Drug Discov Today* 13 (19–20):863–8.

Bar-Lev, S. 2008. "We are here to give you emotional support": Performing emotions in an online HIV/AIDS support group. *Qual Health Res* 18 (4):509–21.

Bender, J. L., L. O'Grady, and A. R. Jadad. 2008. Supporting cancer patients through the continuum of care: A view from the age of social networks and computer-mediated communication. *Curr Oncol* 15 Suppl 2:s107 es42–7.

Bers, M. U., J. Gonzalez-Heydrich, and D. R. DeMaso. 2002. Future of technology to augment patient support in hospitals. *Stud Health Technol Inform* 80:231–44.

Bessell, T. L., S. McDonald, C. A. Silagy, J. N. Anderson, J. E. Hiller, and L. N. Sansom. 2002. Do Internet interventions for consumers cause more harm than good? A systematic review. *Health Expect* 5 (1):28–37.

Black, M., and H. Hussain. 2000. Hydrazine, cancer, the Internet, isoniazid, and the liver. *Ann Intern Med* 133 (11):911–3.

Bos, L., D. Carroll, and A. Marsh. 2008. The impatient patient. *Stud Health Technol Inform* 137:1–13.

Briceño, A. C., M. Gospodarowicz, and A. R. Jadad. 2008. Fighting cancer with the Internet and social networking. *Lancet Oncol* 9 (11):1037–8.

Broom, A. 2005a. Virtually He@lthy: The impact of Internet use on disease experience and the doctor-patient relationship. *Qual Health Res* 13 (3):325–45.

Broom, A. 2005b. The eMale: Prostate cancer, masculinity and online support as a challenge to medical expertise. *Aust Nz Soc* 41 (1):87–104.

Broom, A. 2005c. Medical specialists' accounts of the impact of the Internet on the doctor/patient relationship. *Health* 9 (3):319–38.

Broom, A., and P. Tovey. 2008. The role of the Internet in cancer patients' engagement with complementary and alternative treatments. *Health* 12 (2):139–55.

Buis, L. R. 2008. Emotional and informational support messages in an online hospice support community. *Comput Inform Nurs* 26 (6):358–67.

Cahill, J. 1998. Patient participation—a review of the literature. *J Clin Nurs* 7 (2):119–28.

Cancer Registry of Norway (Kreftregisteret) 2007. Cancer in Norway 2006. Cancer incidence, mortality, survival, and prevalence in Norway.

Cashen, M. S., P. Dykes, and B. Gerber. 2004. eHealth technology and Internet resources: Barriers for vulnerable populations. *J Cardiovasc Nurs* 19 (3):209–14; quiz 215–6.

Chronaki, C. E., D. G. Katehakis, X. C. Zabulis, M. Tsiknakis, and S. C. Orphanoudakis. 1997. WebOnCOLL: Medical collaboration in regional health care networks. *IEEE Trans Inf Technol Biomed* 1 (4):257–69.

Cline, R. J., and K. M. Haynes. 2001. Consumer health information seeking on the Internet: The state of the art. *Health Educ Res* 16 (6):671–92.

Conrad, P. 1992. Medicalization and social control. *Ann R Soc* 18:209–32.

Conrad, P., and J. W. Schneider. 1992. *Deviance and Medicalization. From Badness to Sickness.* Philadelphia: Temple University Press.

Crocco, A. G., M. Villasis-Keever, and A. R. Jadad. 2002. Analysis of cases of harm associated with use of health information on the Internet. *JAMA* 287 (21):2869–71.

Culver, J. D., F. Gerr, and H. Frumkin. 1997. Medical information on the Internet: A study of an electronic bulletin board. *J Gen Intern Med* 12 (8):466–70.

Cummings, J., B. Butler, and R. Kraut. 2002. The quality of online social relationships. *Commun ACM* 45 (7):103–08.

Das, S., L. Girard, T. Green, L. Weitzman, A. Lewis-Bowen, and T. Clark. 2009. Building biomedical Web communities using a semantically aware content management system. *Brief Bioinform* 10 (2):129–38.

Daub, J., P. P. Gardner, J. Tate, D. Ramskold, M. Manske, W. G. Scott, Z. Weinberg, S. Griffiths-Jones, and A. Bateman. 2008. The RNA WikiProject: Community annotation of RNA families. *RNA* 14 (12):2462–4.

Demiris, G. 2006. The diffusion of virtual communities in health care: Concepts and challenges. *Patient Educ Couns* 62 (2):178–88.

Deshpande, A., and A. R. Jadad. 2006. Web 2.0: Could it help move the health system into the 21st century? *J Mens Health Cend* 3 (4):332–36.

Dong, Q., M. A. Urista, and D. Gundrum. 2008. The impact of emotional intelligence, self-esteem, and self-image on romantic communication over MySpace™. *Cyberpsychol Behav* 11 (5):577–78.

Donnelly, L. S., R. L. Shaw, and O. B. van den Akker. 2008. eHealth as a challenge to 'expert' power: A focus group study of Internet use for health information and management. *J R Soc Med* 101 (10):501–6.

Ebbinghouse, C. 1998. Frauds, hoaxes, myths, and chain letters: Or, what's this doing in my e-mail box? *Searcher* 6:50–5.

Esquivel, A., F. Meric-Bernstam, and E. V. Bernstam. 2006. Accuracy and self correction of information received from an Internet breast cancer list: Content analysis. *BMJ* 332 (7547):939–42.

Eysenbach, G. 2008. Medicine 2.0: Social networking, collaboration, participation, apomediation, and openness. *J Med Internet Res* 10 (3):e22.

Eysenbach, G., and C. Kohler. 2002. Does the Internet harm health? Database of adverse events related to the Internet has been set up. *BMJ* 324 (7331):239.

Eysenbach, G., J. Powell, M. Englesakis, C. Rizo, and A. Stern. 2004. Health related virtual communities and electronic support groups: Systematic review of the effects of online peer to peer interactions. *BMJ* 328 (7449): 1166–1170.

Eysenbach, G., and J. E. Till. 2001. Ethical issues in qualitative research on Internet communities. *BMJ* 323 (7321):1103–5.

Falkman, G., M. Gustafsson, M. Jontell, and O. Torgersson. 2008. SOMWeb: A semantic Web-based system for supporting collaboration of distributed medical communities of practice. *J Med Internet Res* 10 (3):e25.

Feste, C., and R. M. Anderson. 1995. Empowerment: From philosophy to practice. *Patient Educ Couns* 26 (1–3):139–44.

Finfgeld, D. L. 2000. Therapeutic groups online: the good, the bad, and the unknown. *Issues Ment Health Nurs* 21 (3):241–55.

Flicker, S., D. Haans, and H. Skinner. 2004. Ethical dilemmas in research on Internet communities. *Qual Health Res* 14 (1):124–34.

Fox, S. and S. Jones. 2009. *The social life of health information: Americans' pursuit of health takes place within a widening network of both online and offline sources.* Report available from http://www.pewinternet.org/Report/2009/8/-the-social-life-of-health-information.asp/

Frost, J. H., and M. P. Massagli. 2008. Social uses of personal health information within PatientsLikeMe, an online patient community: What can happen when patients have access to one another's data. *J Med Internet Res* 10 (3):e15.

Frost, J. H., M. P. Massagli, P. Wicks, and J. Heywood. 2008. How the social Web supports patient experimentation with a new therapy: The demand for patient-controlled and patient-centered informatics. *AMIA Annu Symp Proc* 6: 217–21.

Gilmour, J. A. 2007. Reducing disparities in the access and use of Internet health information. A discussion paper. *Int J Nurs Stud* 44 (7):1270–8.

Ginossar, T. 2008. Online participation: A content analysis of differences in utilization of two online cancer communities by men and women, patients and family members. *Health Commun* 23 (1):1–12.

Gorini, A., A. Gaggioli, C. Vigna, and G. Riva. 2008. A second life for eHealth: Prospects for the use of 3-D virtual worlds in clinical psychology. *J Med Internet Res* 10 (3):e21.

Granovetter, M. 1973. The strength of weak ties. *Am J Social* 78:1360–80.

Grimsmo, A. 1988. Self care and self help—a meeting place for patients and general practitioners [in Norwegian]. *Tidsskr Nor Laegeforen* 108 (29B):2654–6.

Guadagno, R., B. Okdie, and C. Eno. 2008. Who blogs? Personality predictors of blogging. *Comput Human Behav* 24 (5):1993–2004.

Hardey, M. 2002. Personal accounts of illness on the Internet: The story of my illness. *Health* 6 (1):31–46.

Hoffman, J. 2005. Awash in information, patients face a lonely, uncertain road. *The New York Times*, August 14, 2005.

Hughes, B., I. Joshi, and J. Wareham. 2008. Health 2.0 and Medicine 2.0: Tensions and controversies in the field. *J Med Internet Res* 10 (3):e23.

Illich, I. 1976. *Medical Nemesis. The Expropriation of Health.* New York: Pantheon Books.

Jadad, A. R., M. W. Enkin, S. Glouberman, P. Groff, and A. Stern. 2006. Are virtual communities good for our health? *BMJ* 332 (7547):925–6.

Jones, M., and I. Alony. 2008. Scientific commons: Blogs—the new source of data analysis. *J Issues Sci Inf Technol* 5:433–46.

Kim, M.-S., and N. S. Raja. 1991. Verbal aggression and self-disclosure on computer bulletin boards. Paper presented at the Annual Meeting of the International Communication Association, May 23–27, 1991, Chicago, IL.

Kleinman, A. 1980. *Patients and Healers in the Context of Culture.* Berkeley: University of California Press.

Klemm, P., K. Reppert, and L. Visich. 1998. A nontraditional cancer support group. *Internet. Comput Nurs* 16 (1):31–6.

Kovic, I., I. Lulic, and G. Brumini. 2008. Examining the medical blogosphere: An online survey of medical bloggers. *J Med Internet Res* 10 (3):e28.

Lasker, J. N., E. D. Sogolow, and R. R. Sharim. 2005. The role of an online community for people with a rare disease: Content analysis of messages posted on a primary biliary cirrhosis mailinglist. *J Med Internet Res* 7 (1).

Lefer, T. B., M. R. Anderson, A. Fornari, A. Lambert, J. Fletcher, and M. Baquero. 2008. Using Google Earth as an innovative tool for community mapping. *Public Health Rep* 123 (4):474–80.

Leroy, G., E. Eryilmaz, and B. T. Laroya. 2006. Health information text characteristics. *AMIA Annu Symp Proc* 2006: 479–83.

Lupton, D. 1998. Foucault and the medicalisation critique. In *Foucault, Health and Medicine*, edited by A. Petersen and R. Bunton. London: Routledge.

Malik, S. H., and N. S. Coulson. 2008. Computer-mediated infertility support groups: An exploratory study of online experiences. *Patient Education and Counseling* 73 (1):105–13.

McDaniel, E., and P. Duckler. CarePages FAQ. Accessed March 30, 2009. Available from http://cms.carepages.com/export/sites/default/CarePages/en/Press/carepages_faq.pdf.

McKenna, R. J., D. Wellisch, and I. F. Fawzy. 1995. Rehabilitation and supportive care of the cancer patient. In *American Cancer Society Textbook of Clinical Oncology* (2nd ed.), edited by G. Murphy, W. Lawrence, and R. Lenhard Jr. Atlanta, GA: American Cancer Society.

McKinney, A. A., and K. Page. In press. Podcasts and videostreaming: Useful tools to facilitate learning of pathophysiology in undergraduate nurse education? *Nurse Educ Pract.* p. 642.

Mehta, S. A. 2007. What can physicians learn from the blogs of patients with uveitis? *Ocul Immunol Inflamm* 15 (6):421–3.

Moore, T. D., and M. A. Serva. 2007. Understanding member motivation for contributing to different types of virtual communities: A proposed framework. Paper presented at SIGMIS-CPR '07. April 19–21. St. Louis, MO: ACM.

Pennebaker, J. W., and S. K. Beall. 1986. Confronting a traumatic event: Toward an understanding of inhibition and disease. *J Abnorm Psychol* 95 (3):274–81.

Polomano, R. C., N. Droog, M. C. Purinton, and A. S. Cohen. 2007. Social support Web-based resources for patients with chronic pain. *J. Pain Palliat Care Pharmacather* 21 (3):49–55.

Pownall, E. 2004. Using a patient narrative to influence orthopaedic nursing care in fractured hips. *J Orthop Nurs* 8 (3):151–9.

Pratt, W., K. Unruh, A. Civan, and M. Skeels. 2006. Managing health information in your life. *Commun ACM* 49 (1):51–5.

Preece, J., and D. Maloney-Krichmar. 2003. Online communities: Focusing on sociability and usability. In Jacko and A. Sears (Eds) *Handbook of Human-Computer Interaction.* Mahwah, NJ: LEA Press. 596–620.

Rimer, B. K., E. J. Lyons, K. M. Ribisl, J. M. Bowling, C. E. Golin, M. J. Forlenza, and A. Meier. 2005. How new subscribers use cancer-related online mailing lists. *J Med Internet Res* 7 (3):e32.

Sandars, J., M. Homer, G. Pell, and T. Croker. In press. Web 2.0 and social software: The medical student way of e-learning. *Med Teach* 1–5.

Sandaunet, A. G. 2008. *Keeping Up With the New Health Care User: the Case of Online Self-Help Groups for Women with Breast Cancer.* Norway: University of Tromsø.

Sharpton, T. J., and A. A. Jhaveri. 2006. Leveraging the knowledge of our peers: Online communities hold the promise to enhance scientific research. *PLoS Biol* 4 (6):e199.

Smith, C. A., and P. J. Wicks. 2008. PatientsLikeMe: Consumer health vocabulary as a folksonomy. *AMIA Annu Symp Proc* 6: 682–6.

Smyth, J., and R. Helm. 2003. Focused expressive writing as self-help for stress and trauma. *J Clin Psychol* 59 (2):227–35.

Sproull, L., and S. Keisler. 1986. Reducing social context cues; electronic mail in organizational communication. *Manag Sci* 32:1492–512.

Stern, P. N. 1986. *Women, Health, and Culture.* New York: Hemisphere Publishing p. xi–xiv.

Stoddard, J. L., E. M. Augustson, and R. P. Moser. 2008. Effect of adding a virtual community (bulletin board) to smokefree.gov: Randomized controlled trial. *J Med Internet Res* 10 (5):e53.

Tan, L. 2008. Psychotherapy 2.0: MySpace™ blogging as self-therapy. *American Journal of Psychotherapy* 62 (2):143–63.

Tang, P. C., J. S. Ash, D. W. Bates, J. M. Overhage, and D. Z. Sands. 2006. Personal health records: Definitions, benefits, and strategies for overcoming barriers to adoption. *J Am Med Inform Assoc* 13 (2):121–6.

Tangen, L. M., and A. H. Tjora. 2002. Surfing for health: The Internet as health information provider. Paper presented at the BSA Medical Sociology Conference, York, UK.

Thompson, L. A., K. Dawson, R. Ferdig, E. W. Black, J. Boyer, J. Coutts, and N. P. Black. 2008. The intersection of online social networking with medical professionalism. *J Gen Intern Med* 23 (7):954–7.

Tilley, D. S., C. Boswell, and S. Cannon. 2006. Developing and establishing online student learning communities. *Comput Inform Nurs* 24 (3):144–9; quiz 150–1.

Tjora, A. 2008. The modern patient in a sociological light [in Norwegian]. In *The Modern Patient* [in Norwegian], edited by A. Tjora. Oslo, Norway: Gyldendal Akademisk, p. 11–30.

Tsai, C. C., S. H. Tsai, Q. Zeng-Treitler, and B. A. Liang. 2007. Patient-centered consumer health social network Websites: A pilot study of quality of user-generated health information. *AMIA Annu Symp Proc* 11: 1137.

Van Uden-Kraan, C., C. Drossaert, E. Taal, C. Lebrun, K. Drossaersbakker, W. Smit, E. Seydel, and M. Vandelaar. 2008. Coping with somatic illnesses in online support groups: Do the feared disadvantages actually occur? *Comp Human Behav* 24 (2):309–24.

Walther, J. B. 2004. Language and communication technology. Introduction to the special issue. *Journal of Language and Social Psychology* 23 (4):384–96.

Warner, D. A., and J. D. Procaccino. 2004. Toward wellness: Women seeking health information. *J Am Soc Inform Sci Tech* 55 (8):709–30.

Weingart, S. N., D. Rind, Z. Tofias, and D. Z. Sands. 2006. Who uses the patient Internet portal? The PatientSite experience. *J Am Med Inform Assoc* 13 (1):91–5.

Weinman, J., and K. J. Petrie. 1997. Illness perceptions: A new paradigm for psychosomatics? *J Psychosom Res* 42 (2):113–6.

Whitlock, J., W. Lader, and K. Conterio. 2007. The Internet and self-injury: What psychotherapists should know. *J Clin Psychol* 63 (11):1135–43.

Wikipedia Contributors. Semantic Web. Wikipedia, The Free Encyclopedia. Accessed March 30, 2009. Available from http://en.wikipedia.org/wiki/Semantic_Web.

Williams, S. J., and M. Calnan. 1996. The 'limits' of medicalization?: Modern medicine and the lay populace in 'late' modernity. *Soc Sci Med* 42 (12):1609–20.

Woodruff, S. I., C. C. Edwards, T. L. Conway, and S. P. Elliott. 2001. Pilot test of an Internet virtual world chat room for rural teen smokers. *J Adolesc Health* 29 (4):239–43.

Wright, A., D. W. Bates, B. Middleton, T. Hongsermeier, V. Kashyap, S. M. Thomas, and D. F. Sittig. 2009. Creating and sharing clinical decision support content with Web 2.0: Issues and examples. *J Biomed Inform* 42 (2):334–46.

Youngblood, P., L. Hedman, J. Creutzfeld, L. Fellander-Tsai, K. Stengard, K. Hansen, P. Dev, S. Srivastava, L. Kusumoto, A. Hendrick, and W. L. Heinrichs. 2007. Virtual worlds for teaching the new CPR to high school students. *Stud Health Technol Inform* 125:515–9.

Zeng, Q., S. Kogan, N. Ash, R. A. Greenes, and A. A. Boxwala. 2002. Characteristics of consumer terminology for health information retrieval. *Methods Inf Med* 41 (4):289–98.

Zickmund, S. L., R. Hess, C. L. Bryce, K. McTigue, E. Olshansky, K. Fitzgerald, and G. S. Fischer. 2008. Interest in the use of computerized patient portals: Role of the provider-patient relationship. *J Gen Intern Med* 23 Suppl 1:20–6.

Zimmerman, N., C. Patel, and D. P. Chen. 2008. ChMP: A collaborative medical history portal. *AMIA Annu Symp Proc* 859–63.

Zola, Irving K. 1972. Medicine as an institution of social control. *Sociol Rev* 20(4): 487–504.

Chapter 6

Online Support Communities

Ulrike Pfeil

Contents

6.1 Introduction

Online support communities are formed by people who share similar life experiences and who build a space of support, compassion, and trust (Preece and Ghozati, 2001). They offer the possibility for people to interact with others who have similar interests or who face a similar life situation. Harris (2006) describes online support communities as places where people can discuss their feelings and thoughts about illnesses, social and psychological problems, and many kinds of other topics. Online support communities often have the function of self-support groups, which are characterized by a high level of emotional support and understanding.

The investigation of social support in online communities is of great interest to researchers in the area of computer-mediated communication (CMC). This can be seen by the large number of empirical studies on online social support, including email support networks for people with cancer (Rodgers and Chen, 2005; Winefield, 2003, 2006), online forums for people with knee injuries (Preece, 1998), caregivers (Klemm and Wheeler, 2005; White and Dorman, 2000), and many more. Some argue that the formation of social relationships and exchange of emotional support in CMC settings can potentially be very beneficial for its members (Pfeil and Zaphiris, 2007; Preece, 2000; Ellison et al., 2007; Lange, 2007; Ridings and Gefen, 2004; Valkenburg and Peter, 2007; Walther and Boyd, 2002; Eastin and LaRose, 2005). Others disagree and mention disadvantages and limitations of personal and emotional communication in an online environment (e.g., Kraut et al., 1998; Kiesler and Kraut, 1999; Nie, 2001).

It is important to investigate online support communities in order to make an informed judgment about their benefits and problems. We need to know what constitutes social support in online settings and how members perceive it in order to understand how online support communities can be utilized to provide the best possible support for its members.

Different methods have been applied in order to study online support communities. Although some researchers focus on the analysis of communication content (e.g., Preece, 1998, 1999; Coulson, 2005; Rodgers and Chen, 2005; Pfeil and Zaphiris, 2007; White and Dorman, 2000; Wright, 2000a), others are interested in the perception of support by the members of the community (Xie, 2005, 2008; Maloney-Krichmar and Preece, 2005; Kanayama, 2003; Wright, 2000b; McMellon and Schiffman, 2002; Bowker and Tuffin, 2002; Bowker and Tuffin, 2003) or the social network that evolves within online support communities (Wright, 1999; Zaphiris and Sarwar, 2006). For example, Preece and her colleagues explicitly investigated the content of the discussion board postings within an online support community for people with a specific knee injury (Preece, 1998, 1999; Preece and Ghozati, 1998) and studied the usability and sociability of the Web site (Maloney-Krichmar and Preece, 2005). In contrast to that, Wright (1999, 2000b) and Xie (2005) were interested in the impact that supportive communication and the exchange of online support has on people's life and well-being.

When describing the characteristics of online social support and investigating possible benefits and challenges, studies often compare online with offline support. The differences found in support in these settings are often grounded in principal differences between online and offline communication, for example, the lack of nonverbal cues in online communication (Pfeil et al., 2009).

When investigating the perceived benefits and disadvantages of online support communities, researchers often analyze online support communities targeted at a specific user population, for example, older people (Wright, 1999, 2000a, 2000b; Xie, 2007, 2008; Kanayama, 2003; McMellon and Schiffman, 2002; Pfeil and Zaphiris, 2007) or people with a specific disability (Braithwaite et al., 1999; Bowker and Tuffin, 2002; Bowker and Tuffin, 2003). Different user populations have different needs and preferences concerning the use of online support communities, for example, time constraints (as in the case of care providers) or difficulties in using the computer (as in the case of people with disabilities.) It is important to consider these user-specific characteristics to make any conclusions about the applicability and usefulness of online support communities for these specific user groups. In order to investigate the kind of support, as well as the perceived benefit and disadvantages of online support communities, studies need to focus on specific user populations.

This chapter provides an overview over research that has been done in the area of online support communities and discusses a selection of methods that can be applied in order to investigate them. In Section 6.2, I review and reflect on existing literature to define the characteristics of online support. I then proceed to discuss the advantages and disadvantages of online support communities in comparison to offline support (Section 6.3). In Section 6.4, I discuss a number of different methods that researchers currently use in order to study online support communities. Section 6.5 presents a case study to show how different methods can be applied to the same online support community in order to study different aspects of this online support community. It also shows how results of these studies can be integrated to provide a holistic overview over the characteristics of social support as exchanged in this specific online support community. Section 6.6 summarizes the main issues and concludes with directions for further research.

6.2 Social Support

In order to understand how people exchange social support in online settings, we first have to understand how people define social support offline. So far, no consensus about the definition of social support has been reached. Depending on the focus of each study, measurements and components of social support vary widely from study to study. Generally, social support is described as the exchange of verbal as well as nonverbal messages in order to communicate emotional and informational messages that reduce the retriever's stress and "directly or indirectly, communicate to an individual that she or he is valued and cared for by others" (Barnes and Duck, 1994, p. 176). This definition focuses on the content of supportive (verbal or non-verbal) messages. But according to Ford et al. (1996) one has to be careful with just looking at the communication content when investigating the exchange of social support as "[f]irst, not all ostensibly supportive social interactions are experienced as supportive ... Second, the supportee's perception of the quality or substance of social support is a better predictor of successful coping than the sheer number or quantity of support at one's disposal" (p. 189).

In the following, I first focus on theories of social support and discuss how they can be applied to online support communities. Then, I discuss different characteristics of social support as exchanged in online as well as offline settings.

6.2.1 Theories of Social Support

Social support theories aim to explain how the social relationships that we have with others influence our well-being. They explain in detail how our social relationships alter the ways we think, feel, and behave. In addition, they describe how the reception of social support buffers and mediates coping with stressful life events. In their overview over social support theories and measurements, Lakey and Cohen (2000) distinguish between three main theories of social support: The stress and coping perspective, the social constructionist perspective, and the relationship perspective. I will give a brief overview of these three perspectives in the next section and discuss their application and relation to studies investigating online support communities.

6.2.1.1 The Stress and Coping Perspective

The stress and coping perspective claims that the reception of social support acts as a buffer between stressful life events and health. The more social support a person receives, the easier it is for this person to deal with stress and the less this person's health will be affected by stress. This means that social support increases the ability of a person to cope with stress (Lakey and Cohen, 2000). Wright (1999) investigated the relationship between social support, perceived stress, and coping strategies for people participating in online support communities for older people. He found that participating in online support communities encourages their members to take on a coping strategy of "direct action" in order to deal with stress. This way of coping is considered a positive strategy in order to deal with stress. Negative coping strategies like venting about the problem or avoiding it have not been found to be common for people participating in the investigated online support community (Wright, 1999).

According to the stress and coping perspective, it is not only the amount of social support that is received that has an influence on well-being but also the receiver's *perception* of the social support that he or she gets. Not only received, but also perceived support has an influence on the general well-being and health of a person (Lakey and Cohen, 2000). For the perceived support to be high, it is important that the kind of support matchs the need of the person in need of support. Mismatching, meaning that the efforts and actions taken to help are not well matched to the needs of the recipient, can cause difficulties and can have a negative effect on the relationship between supporter and recipient (Goldsmith, 1992). This mismatching is believed to be less prevalent in online support communities for several reasons: Members of online support communities have often had similar experiences and thus understand each other's feelings and thoughts (Pfeil and Zaphiris, 2007). Also, online support communities bring together people with a lot of expertise concerning the relevant topic. As a result, the emotional understanding as well as the factual expertise help ensure that the support retrieved in online support communities matches the needs of the recipient (Maloney-Krichmar and Preece, 2005). However, for the support to be perceived as helpful, it is important that the online support community has a well-defined topic and the members stick to that topic in their communication. Too much off-topic communication or an unspecified purpose of the online support community weakens the emphasis of similarity between members, and thus mismatching of support might occur more often.

6.2.1.2 The Social Constructionist Perspective

The basis of the social constructionist perspective is that people construct their own view of the world, themselves, and their social relations. Thus, people have different constructs and theories about themselves and the social relations that they are engaged in. This suggests that people do not have a clear consensus about what constitutes social support, and some might consider a specific behavior as supportive whereas others might think that this behavior is not at all supportive (Lakey and Cohen, 2000). Therefore, the focus of the social constructionist perspective is mainly on the perception of support. It claims that people develop a stable and consistent view about the supportiveness of others in their life and judge the actions of others in a way that they fit into this view. This means that people who generally have a higher perception of support interpret the behavior of others as more supportive than people with a lower perception of support do, although the objective behavior might be the same. Furthermore, people's perception of support is more influenced by their general impression of how supportive their environment is than by the actual individual action (Lakey and Cohen, 2000).

In online environments, people have fewer cues at hand for the perception of others and thus also for the construction of their social context. Often, the only cue that people have about each other is the text that they exchange. The hyperpersonal perspective (Walther, 1996) states that senders of CMC messages tend to stress similarities to others when presenting themselves in online support communities; and thus people perceive each other and the online support community more positively. Differences and discrepancies between people that might be obvious and salient in offline conversation are hidden by the fact that the communication is mediated by the computer. Online, people can take time to work on their self-representation, and members often arrange their own image in a way to appear friendly and knowledgeable to others (Walther and Boyd, 2002). Thus, positive thinking and a feeling of support and togetherness are often stressed in online support communities and lead to a supportive and positive perception of the online support community by its members. Once an online community is considered supportive by its members, this construct is regularly reinforced by praising the feeling of togetherness and level of support in the online support community (Pfeil and Zaphiris, 2007).

6.2.1.3 The Relationship Perspective

The relationship perspective views social support not as a separate phenomenon but as connected to relationship processes that co-occur with social support (e.g., communication, companionship, low conflict, and intimacy.) This perspective claims that having relationships to others leads to social support as well as health. Isolation—in contrast to being connected in a social network—is believed to be linked to lower self-esteem and having a negative effect on health within this perspective (Lakey and Cohen, 2000).

Several studies have focused on the social relationships between people in online support communities and the social network that these relationships form. For example, Nahm et al. (2003) investigated the correlation of online support community participation and members' relationships to others within the online support community. Through a quantitative analysis of questionnaire results, they developed a structural equation model of social support in CMC for older adults. Results showed that the amount of time that members spend in the online community and computer knowledge correlate positively with the network size of individual members. The more relationships people have with other members within the online support community, the more support they receive. In addition, women were found to have larger networks than men did. Furthermore, difficulties accessing and using the online community due to physical constraints were found to have a negative influence on the network size and the physical well-being of the member. Similarly, Wright (1999) concluded that the amount of time people spend in an online support community has a positive impact on their network size and the satisfaction with the received support.

6.2.2 Characteristics of Social Support

Researchers take many different approaches to describing support, and almost all definitions describe social support as a phenomenon with multidimensional aspects. Whereas some definitions stress informational support (e.g., Cobb, 1976), others focus on tangible support (Craven and Wellman, 1973) or on the basic social needs of people such as approval, esteem, succor, and belonging (Kaplan et al., 1977). Most definitions of social support include tangible components such as financial or practical help as well as intangible components such as encouragement (Heitzmann and Kaplan, 1988).

Similar to offline research, research on social support in online support communities describes the nature of social support in terms of different sets of categories. For example, Moursund (1997) investigated the aspects of the social support exchanged in "the sanctuary," a MUD (multi user domain) for adult survivors of sexual abuse. He distinguishes between expressions of companionship (e.g., talking about problems, sharing experiences), information, positive feedback, motivational support, and belongingness (Moursund, 1997). Similarly, Klemm and Wheeler (2005) studied the messages of an online cancer caregiver listserv over a period of two months. Through inductive content analysis, three major themes emerged from the messages: hope, emotional roller coaster, and physical/emotional/psychological responses. Another commonly used typology is the list by Cutrona and Suhr (1992), who describe social support in terms of informational support, emotional support, esteem support, tangible aid, and social network support. Although this categorization was originally developed for the exchange of social support in offline settings, it was also applied in order to analyze the exchange of social support in online support communities (e.g., Braithwaite et al., 1999; Coulson, 2005; Coulson et al., 2007). These studies found that emotional and informational support is most common in the online support communities that they investigated, and tangible support is, as expected in online environments, least common (Braithwaite et al., 1999; Coulson, 2005; Coulson et al., 2007).

Similarly, Preece and her colleagues did a series of studies investigating empathic messages in online communities. Focusing on an online support community for patients with a specific knee injury, Preece (1998) investigated in detail the communication content. She distinguished between three main types of communication: the exchange of factual information, members writing about their own experiences, and empathic messages that voiced understanding of others' thoughts and feelings (Preece, 1998). She found that almost half of the messages (44.8%) within this support community are classified as empathic (Preece, 1998), 17.4% contained factual information, and in 32.0% of the messages, people wrote about their own experiences.

In summary, categorizing and analyzing the exchanged information in online support communities suggest that people seek online support communities for informational as well as emotional support (Coulson, 2005; Coulson et al., 2007; Braithwaite et al., 1999; Pfeil and Zaphiris, 2007).

6.3 Online versus Offline Support

When it comes to the definition of online social support, researchers often compare the characteristics of online social support with those of its offline counterpart. However, the social support that is exchanged in online support communities is different from social support in offline settings. The unique characteristics of online communication offer benefits but also pose challenges for the exchange of social support. The following section discusses online social support in comparison to offline support. Focus is placed on differences between these two settings concerning group characteristics, relationships between support seeker and support provider, anonymity, the issues of access and availability, and the characteristics of supportive interaction.

6.3.1 Group Characteristics

In order to review online social support, it is helpful to consider how online support communities resemble offline support groups (Braithwaite et al., 1999). Robinson (1988) characterized offline support groups according to several characteristics, many of which are also applicable for online support communities, for example, (i) all members of the group experience or have experienced a common life situation; (ii) group members meet regularly (in case of online support communities, members visit the online support community regularly) and exchange mutual aid; and (iii) those who seek help also provide help to others that is considered beneficial not only for the receiver but also for the support provider. According to Coulson et al. (2007), online support communities have the additional benefit of greater access (members can access the online support community from different locations and at different times), being able to read and write messages at their own pace, and greater confidentiality (members do not meet physically and are often only known by their username) of its members.

Differentiating between group characteristics of online and offline support communities, it is commonly argued that the exchange of social support in online support communities facilitates weak ties rather than strong ties between its members (Wellman and Gulia, 1999). Social support in offline settings is usually exchanged via strong ties (e.g., among family members, friends, work colleagues.) Often, these relationships are multiplex and evolve around a number of topics (Leatham and Duck, 1990). The relationships between people in online support communities, in contrast, are usually based on one common experience and thus more likely to be uniplex. Often, there are more people available through weak ties than through strong ties. In addition, weak ties are usually more diverse than the homogeneous group of strong ties and thus offer a wider spectrum of knowledge and expertise. Online support communities resemble such a network of weak ties and provide access to a diverse pool of experienced people. Often, people turn to online support communities because they could not find the kind of support that they need in their immediate network of close ties (Walther and Boyd, 2002; Winefield et al.,

2003; Buchanan and Coulson, 2007). Also, the greater distance between people connected via weak ties is considered beneficial. For example, Adelman et al. (1987) state, "This network distance enhances perceived anonymity and allows people to seek information and support without having to deal with the uncertainty of how those in primary relationships might respond" (p. 131). In this way, they argue, weak links facilitate "low-risk discussions about high-risk topics" (p. 133).

6.3.2 The Relationship between Support Giver and Support Seeker

The fact that social support is exchanged in online settings also has an impact on the relationship between the support seeker and the support giver. In offline settings, people usually exchange social support after they have established a personal relationship with the other person. In online settings, people often come straight to the topic of concern and reveal a great deal of personal and emotional information without establishing relationships with the other members (Walther and Boyd, 2002). However, although people talk to relative strangers in online support communities, studies have shown that they often develop a high level of trust within the online support community (Pfeil and Zaphiris, 2007). The development and maintenance of trust is hereby often described to be more difficult and time consuming but nonetheless possible and beneficial for online support community members (Walther, 1996).

In face-to-face interactions, the support seeker has to openly admit weaknesses and ask for help from the support giver, which often leads to the support seeker being considered less competent and sometimes even stigmatized according to the information that she or he discloses (Albrecht et al., 1994, p. 433). The fact that the support seeker admits feelings of fear or incompetence and openly asks for help creates a situation of dependency of the support seeker on the support giver (La Gaipa, 1990, p. 136). Sometimes this leads to a rejection of the help given in order to restore the balance within the relationship. In addition, seeking support from others in offline situations often creates the expectancy that the support seeker will reciprocate the service. A denial of such reciprocation would influence equity and balance in the relationship and eventually harm it.

In contrast to that, social support in online support communities is exchanged between people who otherwise are not related to each other. This ameliorates the possible inequalities and uncertainties that come up when a person discloses personal information that weakens her or his position. Also, the expectation of reciprocity as it occurs in offline situations is less likely to be an issue in online support communities. One can just decide to quit posting and leave the online support communities without having to return the favor. Often, this is not even recognized by the other members of the online support communities, as there are usually many supportive messages from other members.

Preece and Ghozati (2001) also state that similarity of members is very important for the exchange of support in online settings. People who have experienced a similar situation are more likely to empathize with each other, as they can understand the complexity of the situation much better than outsiders can (Ickes, 1997). Also, similarity is beneficial for the development of trust in online support communities, as people who are similar tend to be less suspicious of each other (Wallace, 1999).

6.3.3 The Influence of Anonymity

The fact that members of online support communities are anonymous greatly influences the exchange of social support, as people feel safer to disclose personal information when they do not have to disclose their identity. Thus, anonymity might be one of the most important reasons people exchange deeply personal and emotional information in online support communities (Walther and Boyd, 2002). People feel more open to talk about personal and emotional details of their lives in a safe environment where disclosing personal information and problems is common and expected. In addition, anonymity also prevents members from being judged according to their gender, race, and age (Wallace, 1999).

However, anonymity can also have a negative influence on communication in online support communities, as people are more prone to misbehave and post irresponsible and hostile messages in online support communities. Preece and Ghozati (2001) point out that the lack of negative consequences for hostile messages might be the reason for misbehavior. Despite this, their study showed that online support communities are very unlikely to contain hostile messages but instead contain more empathic messages than do other kinds of online communities. In addition, moderation of online communities as well as the presence of women in these online communities increases the likelihood of empathic messages (Preece and Ghozati, 2001).

Being anonymous also makes faking illnesses very easy in online support communities, and people might pretend to suffer from a disease. Although this behavior is not openly hostile, it can have serious consequences for the trust within the online support community (Preece and Ghozati, 2001), especially for a vulnerable user group (Bowker and Tuffin, 2003).

6.3.4 Access and Availability

For many people, online support communities are easier to access than are offline support groups, thus enabling them to get support that would not be available to them offline. The fact that there are no temporal and geographic restrictions for members to participate in online support communities allows them to utilize the online support communities according to their pace and without having to leave the house (Braithwaite et al., 1999). Also, people can write and read messages whenever they want to and can put in as much time and effort as they wish to (Coulson et al., 2007). In addition, people can concentrate on what they want to write in a message without having to listen to the other person at the same time. This might in some cases enhance the quality of the social support, as more time and effort can be put into a supportive reply (Walther and Boyd, 2002).

In addition to the issue of access to the support community itself, a unique characteristic of online support communities is that they often have an archive in which all the messages are stored. This enables members to read these messages at all times, even if the original posters are no longer present, a feature unavailable in offline groups. Having access to a large number of messages offers the opportunity for online support community members to compare themselves with other members and conduct a form of "reality check." Finding similarities helps members realize that they are not alone with their thoughts and feelings and attribute normality to their behaviors. The comparison with a variety of people especially enables them to judge how normal or typical their own thoughts and feelings are (Adelman et al., 1987). This is possible not only for people who actively post messages in the online support community but also for so-called "lurkers"-members who only read but do not post messages (Walther and Boyd, 2002). As Mickelson (1997) states, lurkers in online support communities "can obtain comparison information or vicarious support without having to disclose anything about themselves...[and] obtain validation for their feelings of stigma without having to communicate those feelings to others" (p. 172).

6.3.5 Characteristics of Supportive Interaction

The way people exchange support in online support communities is unique. Instead of explicitly asking for support as is often done in offline settings, support seekers in online support communities disclose their thoughts and feelings in the form of a personal narrative and then ask if others have had similar experiences. This kind of message then triggers supportive responses from other members within the online support community (Pfeil and Zaphiris, 2007). Also, online support community members tend to post questions like "What would happen if..." in order to explore possible effects of a certain behavior or decision and encourage others to talk about their experiences.

When giving support, studies found that members of online support communities tend to disguise advice and help in talking about their own experiences. Rather than giving explicit suggestions to the other person on how to behave in a certain situation, they talk about how they dealt with the same or a similar issue (Wright, 2000a; Pfeil and Zaphiris, 2007). As members in online support communities often experience a common life situation, support seekers usually find people who have experienced a similar situation and thus can give valuable information and emotional support by sharing their experiences.

6.4 Analyzing Online Support Communities

The following sections will discuss different methods that can be applied to study online support communities. In particular, I will elaborate on content analysis, the analysis of interactivity and responsiveness, social network analysis (SNA), and query-based techniques. The main aspects of the different methods will be presented, and examples of how they can be applied in order to study online support communities will be given.

6.4.1 Content Analysis

Content analysis is one of the most commonly applied methods for investigating online support communities. It facilitates analysis by describing the content of online supportive communication in terms of a set of categories or themes. Studies that apply content analysis to investigate online support communities usually look at a subset of the messages. In some studies, whole messages are sorted into categories (e.g., Preece and Ghozati, 1998; Coulson et al., 2007), whereas others divide messages into subunits and sort these into categories (e.g., Pfeil and Zaphiris, 2007). These categories are then further analyzed, either in terms of qualitative descriptions or quantitatively (Mayring, 2000; Krippendorff, 1980). Quantitative content analysis emphasizes the statistical analysis of the data (e.g., test hypothesis with quantified measurements), whereas qualitative content analysis focuses on the themes and topics of the categories and their meanings and relations between them. Frequency calculations can be part of the qualitative content analysis as well but are used less rigorously than in quantitative content analysis (Sandelowski, 2000).

Depending on the aim of the study, researchers either apply inductive content analysis with the aim to develop a framework of a set of categories (e.g., Pfeil and Zaphiris, 2007; Rodgers and Chen, 2005; Preece, 1999), or they follow the deductive approach, sorting the messages into a predefined set of categories (e.g., Braithwaite et al., 1999; Coulson et al., 2007; Preece and Ghozati, 2001; Winefield, 2006).

The deductive approach is based on a predeveloped set of categories and imposes this framework on the data. Often, the set of categories that is used describes support in general (e.g., Cutrona and Suhr, 1992) and is rarely tailored to a specific kind of target population or media in which support is communicated. However, these sets are often verified by multiple applications in different research studies. For example, Cutrona and Suhr's (1992) framework of support was applied to investigate online support communities in several studies (e.g., Braithwaite et al., 1999; Coulson, 2005; Coulson et al., 2007). Thereby, the number of occurrences of the different categories is often compared to other studies, or one study investigates different online support communities and then compares the frequency of categories among them. For example, Preece and Ghozati (2001) compared the level of empathy and hostility among different kinds of online communities. The advantage of a deductive content analysis is that it allows for a comparison of social support among different settings and thus makes it possible to generalize findings. However, it is important that the set of categories is suitable for investigating the online support community, as the disadvantage of the deductive approach is that the framework is imposed on the communication content although it might not accurately reflect the nature of the communication.

In contrast to the deductive approach, inductive category development constructs a framework that describes the occurring communication patterns (Mayring, 2000). Often, the developed framework is analyzed qualitatively and provides the reader with a description of the communication content (e.g., Rodgers and Chen, 2005). Also, some studies provide the frequencies of the categories to give an idea about the distribution of the different kinds of support within the online support community (e.g., Preece, 1999; Pfeil and Zaphiris, 2007). The benefit of this approach is that the results are not biased by a predefined set of categories, and the developed framework exactly describes the communication patterns of the online support community. The challenge of this approach is, however, that it is difficult to compare findings and discuss them in context with other studies as the set of categories is unique and specifically tailored to the online support community under investigation. Although this kind of empirical investigation is a good way to explore the communication patterns within an online support community, it is difficult to generalize the findings of such an investigation.

In summary, content analysis is applied in order to describe the kind of communication in online support communities. It is a suitable method to conceptualize the content of the communication. However, content analysis is not suitable to find out how people perceive the exchanged support. In addition, it is also unsuitable for investigating how relationships develop in online support communities, as it looks at the content independently from the authors. Further analysis is necessary in order to understand how members perceive this content and how it facilitates the development of relationships between them.

6.4.2 Analysis of Interactivity and Responsiveness

Responsiveness and interactivity are important characteristics of online communities, as responsive and interactive online communities are believed to be more engaging and beneficial for their members (Kalman et al., 2006). In order to investigate the interactivity and responsiveness of messages within an online support community, the key is to identify how messages are related to each other (e.g., whether one message responds to a previous one.) The frequency of responsive messages (messages that respond to a previous one) or interactive messages (messages that comment on how a previous message responded to another message) is calculated and interpreted (Rafaeli and Sudweeks, 1997; Kalman et al., 2006; Jones et al., 2004).

However, when analyzing responsiveness and interactivity in online settings, one has to consider that several conversations can go on at the same time, resulting in related messages not being posted in consecutive order (Greenfield and

Subrahmanyam, 2003; Lapadat, 2002; O'Neill and Martin, 2003). Also, it is quite common in CMC settings that one initial message triggers multiple responses, and one message can refer to multiple other messages. This occurs especially in asynchronous CMC settings where messages are often longer and one message can refer to multiple conversations. Thus, it is important to carefully identify the relations between messages within the online support community.

In addition to the calculation of the frequency of responsive and interactive messages, it can also be beneficial to investigate the content of messages that refer to each other to identify common communication patterns within the online support community. For example, Joyce and Kraut (2006) found that long initial posts or posts that included a question were more likely to trigger a response. Also, responding messages are reported to be similar in style and form to the initial post (Becker-Beck et al., 2005), e.g., initial messages that sound negative trigger a more negative response and longer initial posts trigger longer replies (Joyce and Kraut, 2006). However, these findings are based on the investigation of generic online communities with a mixture of topics. Findings from Fisher et al. (2006) as well as Joyce and Kraut (2006) suggest that the likelihood of responding to initial messages is dependent on the kind of online community. Both studies suggest that online support communities have one of the highest responsiveness scores and thus are most likely to be sustained over a longer period.

The analysis of interactivity and responsiveness of online support communities can give valuable insight into the supportive communication patterns in the investigated setting. It can be combined with other methods investigating the communication behavior of online support community members (e.g., content analysis).

6.4.3 SNA

SNA has been widely used to study offline social networks and has recently been applied to investigate online communities (Kavanaugh and Patterson, 2001; Hampton and Wellman, 2000; Haythornthwaite, 2000; Aviv et al., 2003; Laghos and Zaphiris, 2006; Zaphiris and Sarwar, 2006). SNA is different from standard CMC methods, which often study computer-mediated relations separately from the network in which they occur. Social networks consist of nodes that are linked to each other. When investigating online support communities, nodes represent the members and the links between the nodes describe a connection (e.g., friendship or communication) between these members. The strength of SNA lies in explaining social relations with the structure and patterns of the network in which these relations develop. SNA therefore shifts the focus from individualism to structural analysis of the whole network (Garton et al., 1997).

SNA can be applied quantitatively, measuring characteristics of the network such as density, cohesiveness, etc. These measurements are then interpreted. For example, the more dense a network is, the more interconnected the members of the network are and the more the members of the network have direct contact with each other (Garton et al., 1997). Similarly, in a network with high reciprocity, people tend to respond to each other often and the relations between them are bilateral, whereas in a network with low reciprocity, more one-directional ties exist and the relations are more unbalanced. In addition to the identification of network measurements, SNA can also be applied qualitatively, focusing on the interpretation of network visualizations. The use of graphical representations of the network helps to identify people who are central or isolated in the social network and spot asymmetries in the network structure (Scott, 2000).

Up till now, SNA has rarely been applied in order to study online support communities. Several studies, however, have investigated concepts that are related to those of SNA. For example, Maloney-Krichmar and Preece (2005) investigated communication patterns within an online support community. They found that members within the online support community were densely interconnected via the exchange of messages. Furthermore, they distinguished between key members, who frequently post messages over a long period, community members, who show less activity on the online support community, and lurkers, community members who only read but never write any messages. Based on the characteristics of the relationships between members, they conclude that some relationships resemble strong ties, especially within subgroups of the online support community (Maloney-Krichmar and Preece, 2005). Similarly, Fisher et al. (2006) applied SNA in order to investigate roles in Usenet newsgroups. The individual's social networks and connections to other people helped them to characterize social roles that members take on in online communities.

Furthermore, measurements of an individual's social network size (e.g., the number of friends) have been put into relation to the quality and perception of the exchanged support. For example, Nahm et al. (2003) developed a structural equation model of social support in CMC for older adults. Results showed that time in the online community and computer knowledge correlate positively with the network size, whereas network size decreases with increasing age. The larger the network size, the more support the members received. Furthermore, difficulties accessing and using the online community due to physical constraints were found to have a negative influence on the network size and the physical well-being of the member. Similarly, Wright (1999) found that the amount of time older people spend in an online support community has a positive impact on their network size and their satisfaction with the received support. Older people who spend less time communicating with online community members rated their satisfaction with support from their offline network higher. Eastin and LaRose (2005) developed a model of social support seeking based on questionnaire results from support seekers in online support communities. Their results show that the more socially efficacious people are online and the more active they are in seeking support, the higher the number of their online social network. Also, the size of the online social network is positively related to the perception of the received social support. They conclude that "the more socially efficacious people are online the more likely they are to view the Internet as an important support outlet, spend time seeking support and finally increase the number of people in their online support network" (Eastin and LaRose, 2005, p.989).

As discussed above, current studies mainly investigate the size of an individual's network (number of friends) rather than the characteristics of the whole network. I believe that the additional analysis of relational properties between members in online support communities may give further insight into the impact of online social support on relationship development. SNA can be based both on members reporting the number and names of their friends (resulting in a social network describing friendship) and on the communication activity (e.g., who talks to whom.) Also, SNA can be combined with the analysis of the content of an online support community. For example, it is possible to distinguish between different kinds of communication and investigate the social network of supportive communication opposed to the social network of hostile communication within the same online community.

But when the communication activities within an online support community are examined, it needs to be considered that the findings might not reflect the perception of the members themselves. Although, one member might appear as very popular within the online support community because he or she sends many messages, he or she might not necessarily be considered popular by other online support community members, as the content of the messages might not be perceived as useful. Thus, SNA is a good method to investigate communication behavior and relationships within an online support community, but query-based techniques are needed in order to investigate how the behavior is perceived by the online support community members.

6.4.4 Query-Based Techniques

Query-based techniques have been applied to investigate how members of online support communities perceive the communication activities and the impact of the exchanged support on their daily life. In particular, interviews and questionnaires are commonly used in order to investigate online support communities (e.g., Maloney-Krichmar and Preece, 2005; Wright, 2000a, 2000b; Xie, 2005, 2008).

6.4.4.1 Interviews

Interviews with online support community members are often conducted to investigate the benefits and challenges of online social support as well as how members of online support communities perceive the communication and the support that is being exchanged in the online settings. Also, they are applied in order to elicit the motivation and reasons for members' behaviors. The data collected come straight from the members of the online communities, whereby they report their own personal experiences, activities, and thoughts.

Some studies apply interviews in order to investigate how members of online support communities perceive the exchange of social support (e.g., Xie, 2005, 2008).

Thereby, different aspects and themes of social support in the online support community can be found and described. For example, Xie (2008) distinguishes among informational, emotional, and companionship support, and the results of her interviews show that certain kinds of support are more or less frequent depending on the medium that is used for the communication (e.g., discussion board, instant messaging.) In addition to the analysis of characteristics of support, some studies focus on other aspects of online communication in online support communities. For example, Bowker and Tuffin (2002, 2003) applied interviews to investigate an online community for people with disabilities. In their study, they focused on the role of anonymity and deception (Bowker and Tuffin, 2003) and how people with disabilities choose to disclose information about themselves and their disability (Bowker and Tuffin, 2002). Their findings show in detail how people with disabilities perceive and handle these aspects and what role they play in the exchange of social support online.

Interviews can also be conducted to complement other methods of investigation. For example, Maloney-Krichmar and Preece (2005) applied multiple methods (e.g., content analysis, interaction process analysis, and role analysis) and interviews in order to study in detail the exchange of social support in an online community. In their analysis, they linked all their findings and provided a comprehensive and holistic description of the online support community (Maloney-Krichmar and Preece, 2005).

Interviews are mostly analyzed qualitatively, eliciting key themes and topics that are mentioned by participants of the study. These key topics are then analyzed, for example, to explain the members' perception of the social support. Interviews can give great insight into how members perceive the social support exchanged in online support communities and what they consider the benefits and challenges of the online social support. However, it is important to note that interview findings present people's viewpoints, which are not necessarily reflected in their behavior. Thus, it is useful to combine interviews with other research methods that look at the behavior of members of online support communities (e.g., content analysis or SNA; Maloney-Krichmar and Preece, 2005).

6.4.4.2 Questionnaires

Like interviews, questionnaires are an important technique for collecting members' opinions and experiences with online support communities. Whereas interview questions can be adapted as the interview progresses, questionnaires have a fixed set of questions that cannot be adjusted according to participants' comments during the data collection. Depending on the questions, questionnaires can be analyzed qualitatively as well as quantitatively.

Wright (2000b) conducted an online questionnaire to investigate social support in SeniorNet, an online community for older people. Results show the relation between the amount of time people spend in the online community and their satisfaction with the community support. He found that older adults who spend more time in the online community are more satisfied with the support they receive from the online community (Wright, 2000b). In a second study, Wright (2000a) combined questionnaire and content analysis in order to investigate how older people give and receive social support within SeniorNet. He studied a sample of messages from twenty forums within SeniorNet. Additionally, an online survey was conducted to gather further information about the kind of support that participants experienced within this online community. The themes developed out of the studied messages were promoting community support, advice disguised as self-disclosure, and shared life events. Results from the survey showed that the spectrum of experienced support ranged from information support to highly emotional attachment (Wright, 2000a).

Similar to interviews, questionnaires are a useful method to investigate people's perceptions and opinions but do not consider whether their opinions are actually reflected in their behavior. Thus, as with interviews, it might be beneficial to combine questionnaires with methods that investigate the exhibited behavior specifically.

6.5 Case Study—Social Support in an Online Community for Older People

The following section puts the discussed issues into practice and describes a case study in which social support within an online support community for older people was investigated. I briefly present several studies in which we investigated different aspects of online support communities for older people. I further discuss how the findings can be integrated to provide a holistic description of social support in the investigated setting.

6.5.1 Why Older People?

In recent years, the degree of Internet usage by people aged sixty-five and older has increased significantly (47% between 2000 and 2004). Currently, 28% of older British people go online (Office of Communication, 2006). Similar numbers can be found in the United States, where 22% of American older people use the Internet and this figure is expected to continue to grow (Fox, 2004). Although email is currently the most prevalent communication activity of older people online, the use of online communities is also growing for this target group.

Research has shown that online communication enhances older people's quality of life and well-being (Xie, 2007), as it increases social interaction (Bradley and Poppen, 2003), provides the opportunity to connect to like-minded people (McMellon and Schiffman, 2002), and empowers older people to not only receive but also provide support. Especially for older people who suffer from a specific illness and/or are housebound, online support communities provide the opportunity to meet people who are in a similar situation and to engage in satisfactory social interaction (McMellon and Schiffman, 2002), which can prevent isolation and decrease loneliness (Bradley and Poppen, 2003). Research investigating why older people use CMC found that the opportunity to socially interact with like-minded people and the exchange of social support and companionship are strong motivators (Kanayama, 2003; Pfeil and Zaphiris, 2007). However, there are also special challenges when it comes to online communities for older people, as deception and misbehavior in online support communities can have harmful consequences, especially for a vulnerable user group (Bowker and Tuffin, 2003).

Several studies have investigated the content that older people share in online support communities (Brennan et al., 1991; White and Dorman, 2000; Wright, 2000a; Xie, 2005; Kanayama, 2003). Others have applied query-based techniques to investigate the perception of online social interactions among older people (Wright, 1999; Wright, 2000b; McMellon and Schiffman, 2002) or looked at online social networks (Zaphiris and Sarwar, 2006). Findings showed that older people develop friendships in online communities (Pfeil and Zaphiris, 2007; Xie, 2005), and the social networks that emerge in online communities help them to cope with stressful life situations (Wright, 1999). In a comparison between the social networks of newsgroups for younger and older people, Zaphiris and Sarwar (2006) found that the newsgroup for older people had more consistencies and stability in activity and behaviors of its participants.

However, only few of these studies investigated social support specifically and used multiple methods to conduct their research. An integrated investigation is needed to fully understand how older people exchange social support in online support communities. The following sections describe different substudies that have been conducted to investigate social support in an online support community for older people and discuss how these studies can be integrated in order to provide a comprehensive description of social support in online support communities for older people.

6.5.2 Content of Communication

In order to study the components of social support in online communities for older people, inductive content analysis of a subset of the messages of the discussion board about depression within SeniorNet was conducted. The conversation on the board for a period of 1.5 years between 2000 and 2001 was collected and analyzed. Four hundred messages were exchanged during this time, and qualitative content analysis was used to determine how social support is expressed and facilitated in online communication. The findings identified different components of social support and elicited the different roles that people take on in the online support community (Pfeil and Zaphiris, 2007).

As a result of this study, a code scheme that describes the different aspects of support in the investigated online community was developed. Table 6.1 lists the seven main categories of the code scheme with the short descriptions and examples based on the analysis of the messages. The code scheme describes the characteristics

Table 6.1 Developed Code Scheme

Category	Description	Examples
Self-disclosure	In these text units, people post information about themselves. This can be done in different ways (e.g., emotional, narrative, medical).	"I yawn all the time. I want to go to bed. I know you're supposed to get out, but I don't have the energy to do that much."
Community building	The text unit includes people's opinion about the online community and metainformation about communication activities on the discussion board.	"Thank God for this board, as I can sit here and cry and rattle on—you are the only ones who understand."
Deep support	Supporting text units are often emotional and customized toward the unique situation of the target that the message is for.	"Words are so hard right now. So I place my hand gently over yours and let love and sweetness flow through to you."
Light support	The text unit is supportive and uplifting. It is written in a generic way for another person or the whole community.	"Hang in there; I'm thinking about you."
Factual information	These text units include questions and answers about factual information within the topic (e.g., medication).	"So in 'both cases,' situational depression and bipolar depression, they alter chemicals in the brain?"
Technical issues	The text units are concerned with technical problems or suggestions to solve them.	"Read in your browser screen and have Notepad or Wordpad minimized…"
Off topic	Text units that are about others or about topics that strayed away from the theme of the discussion board.	"Sorry to hear Iowa's weather yesterday. Minnesota is much too cold and damp."

and components of social support as exchanged in online support communities for older people (Pfeil and Zaphiris, 2007).

The nature of the codes, their relationships, and their impact on the communication (as further discussed in Pfeil and Zaphiris, 2007) help us see what makes a successful online support community. By revealing the various characteristics of social support in the investigated online community, we were able to investigate in depth not only the components but also the relations and dependencies necessary to nurture this online support community.

6.5.3 Communication Relationships

In order to study the communication patterns, SNA was applied to the same data set that was used for the content analysis. We investigated who was talking to whom in the online support community and constructed a social network based on the communication activities. In addition to looking at the structure of the exchanged messages within the discussion board as a whole, we investigated the impact of the communication content on the social network patterns. In particular, we investigated whether conversations in each of the seven identified categories (see Table 6.1) have an impact on network characteristics (e.g., density of the network, building of cliques, and inclusiveness of the network.) Figure 6.1 shows the

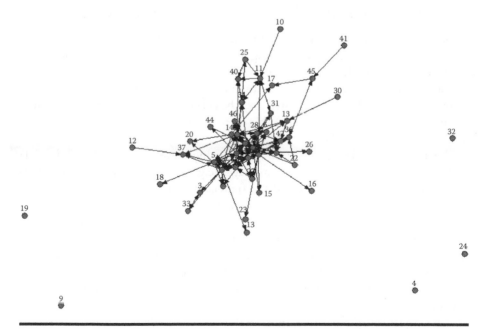

Figure 6.1 Sociogram of the investigated communication activities within the online support community.

sociogram of all investigated communication content of the discussion board about depression within SeniorNet for the period of 1.5 years between August 2000 and February 2001.

Findings show distinct differences between the social network patterns of supportive and nonsupportive communications. For example, members are more connected and closer to each other in the social networks that are related to support than in social networks that are not related to support. Also, the difference between seeking and giving support has an impact on the network structure, as messages that seek support are directed to the whole online support community and messages that give support are more commonly targeted to specific members. Additionally, the results show that the type of support has an impact on the social network structure within the discussion board. Whereas *light support* is freely shared between all members of the online support group equally, *deep support* is often exchanged in small subgroups within the online support community.

The findings elicit the social network structure of the online support community for older people and identify links between network patterns and characteristics of support. They show that, depending on the topic and content of the communication, different structures of social networks emerge. This sheds light on the connection between the kind of communication in online support communities and the relationships that exist between the communicators.

6.5.4 Role of Interactivity in Development over Time

In order to analyze the conversational patterns of the online support community, we investigated interactivity by looking at responsive messages within the online support community. In addition, we were also curious how certain sequences of messages within the online support community are related to the level of activity. We analyzed a data set of messages posted over the period of six years within SeniorNet's discussion board about depression. (This data set includes the 1.5 years investigated in the previous parts of the case study.)

First, we coded the messages into the seven high-level categories (see Table 6.1) and coded the relationship between messages (e.g., one message responds to a previous message.) Then we calculated the frequency of two categories occurring in related messages (using Jeong's, 2005 discussion analysis tool.) A transition probability matrix was calculated that contains the probability that the first category will be followed by the second and investigated whether this sequence occurred significantly more or less often than the random value based on the probability of the individual codes (Jeong, 2005). Also, we investigated whether certain sequences of messages occurred at specific times (e.g., during very busy or quiet periods within the six years.)

The sequence *self-disclosure—deep support—community building* was found to be the basis of communication within the online support community, as it would occur equally in all stages of the evolution of the online community and would often start a conversation after a quiet period. In addition, we also identified message sequences that only occur when the level of activity within the online community is high (*light support—light support* and *technical issues—technical issues*), indicating that these message sequences are only occurring when there is enough other communication going on at the same time. The exchange of factual information, however, showed a clearly different relationship to the level of activity within the online community, as it seems to be independent of the level of activity within the online community and is exchanged when needed. The message sequence *off topic—off topic* was found to be related to a decreasing activity level within the online community, as it only occurred at times of relatively low message frequency and at times when the message frequency is decreasing.

These findings contribute to the frequently discussed topic of how successful online support communities are. The characteristics of the message sequences and their relation to the level of activity within the online support community help us to see what makes a successful and popular community. Also, we identified the role of message sequences for the sustainability of the investigated online support community for older people. By revealing the various characteristics of conversation, we investigated in depth the components that are necessary to nurture an online support community.

6.5.5 User's Perception of Online Social Support

In order to get a deeper understanding about the needs and preferences of older people regarding support in online support communities, we conducted interviews with thirty-one older people (Pfeil et al., 2009). We studied older people's perceptions and experiences of support in their offline lives and their motivation and reluctance to exchange support via online communication (e.g., emails and online support communities.) We did this by conducting detailed interviews with three groups of older people who had different levels of expertise in using the Internet and online communication (non-Internet users, users who use email, and users who use online support communities.) Again, we built on our former findings, and the developed categories (see Table 6.1) were used as a basis for the interviews, and people were asked about their experience of online and offline support relating to these categories. Interviews were transcribed verbatim and analyzed. By comparing and contrasting the perceptions and experiences of the three groups of users, we investigated how the different characteristics of online and offline communication can facilitate or hamper the exchange of support for older people.

Our findings show that online support communities for older people do indeed have the opportunity to enhance their lives, but our findings also indicate that in order to facilitate online support for older people, special care has to be taken to address the needs and preferences of this target group. In this study, we not only elicited information from older people already using online support communities but also integrated the views from older people who have the knowledge and ability to use online support communities but do not want to use them at the moment and older people who are not using the Internet at all. By covering a range of older people with different experiences concerning Internet usage and online support, we provide a widespread overview of their experiences and expectations concerning offline and online support. Including the view of older people who do not use the Internet and do not participate in online support communities, we identified the reasons for reluctance to do so. These findings are very important in spotting problems of online support communities for older people. For example, our findings show that the exchange of *deep support* in online support communities can be highly beneficial for our target population. At the same time, our participants voiced the danger that the exchange of *deep support* can easily lead to misunderstandings in online settings. This knowledge helps us to understand and interpret the way in which *deep support* is exchanged in online support communities for older adults.

Our study investigated the holistic experience of support of older people and the role online support may or may not play in this experience. By putting the experiences of online support in relation to offline support experiences, we did not look at support in online settings in isolation but in relation to offline support. Thus, we elicited valuable information about where expectations, needs, and preferences of older people concerning online support are rooted. These findings help us understand older adults' behavior in online support community settings and in some cases the reasons for their nonparticipation (Pfeil et al., 2009).

6.5.6 Integration of Findings

The last sections present different ways of studying an online support community for older people. In an initial study, we developed a set of categories that describe the supportive communication behavior of older people in the online support community (Section 6.4.2). The subsequent studies built on this code scheme and investigated the categories from different angles. This was done in order to verify the code scheme as well as to deepen our understanding and the description of its categories. The use of content analysis helped us to conceptualize the kind of support that people exchanged in the online support community (Section 6.4.3), and we could further deepen this investigation by studying how this content facilitated relationships between members and how people used individual communication parts to engage in an interactive conversation (Section 6.4.4). Also, the categories were used in order to study how certain content engages or hinders communication activity within the online support community (Section 6.4.4). In addition, by looking at the behavior of members only, we also investigated the perception of online supportive communication by older people and the perceived benefits and challenges of the identified aspects of online supportive communication (Section 6.4.5). Thus, the knowledge and description of the code scheme is multifaceted and can be taken as a framework for studying online support communities for older people.

This case study showed how method triangulation can be applied in order to study the exchange of social support in an online support community for older people. Investigating the exchange of social support from different angles helped us to develop a holistic description of social support in the online support community.

6.6 Summary and Outlook

In this chapter, I first presented a description of social support in general and discussed theories as well as characteristics of social support in offline settings. An identified deficiency is that CMC studies frequently refer to the characteristics of social support (e.g., White and Dorman, 2000; Braithwaite et al., 1999; Coulson, 2005; Coulson et al., 2007); they rarely link their findings to social support theories. In order to gain a comprehensive understanding of online social support, it may be beneficial for further research to also link their empirical findings to social support theories. This would allow a generalization and comparison between empirical studies that currently stand alone.

Also, I described online support in comparison to offline support. When referring to online support, most of the literature was found to investigate social support as exchanged in asynchronous online support communities. However, there are many more ways of communicating support online (e.g., chat and social networking sites.) For example, Xie (2008) investigated how support differs depending on the environment that it is exchanged in. She found differences between support in discussion boards, chat, and instant messaging. I believe further research into the differences and commonalities of social support in different online settings is necessary in order to keep up with the changing technologies.

In addition, I discussed different methods that can be applied in order to study online support communities. Even though these are the most important methods, many more can be used and can give valuable insight into online support communities. For example, the analysis of social roles that members take on in online support communities may give us further insights into members' behavior patterns. There have been some studies that have investigated social roles in online communities. For example, Fisher et al. (2006) identified a set of social roles that people take on in Usenet newsgroups. I believe that the identification of roles in online support communities can help us understand people's behavior in online support communities as well as their motivation for participation. Thus, further research into roles in online support communities is necessary. Similarly, longitudinal studies that investigate the development of online support communities over time are necessary in order to understand the main factors of the sustainability of online support communities.

Finally, I showed how different methods can be combined and method triangulation can be applied in order to study social support in online support communities for older people using a case study as an example. I argued that method triangulation is necessary in order to provide a holistic description of the online support community. Through focusing on a specific target population (e.g., older people), it was possible to identify communication patterns and perceptions unique to this target population. It has been found to be important to consider the target population when investigating online support communities, as different people have different expectations and needs concerning online social support. I would thus encourage further research into online support communities that focuses on a specific user population, e.g., people with disabilities.

References

Adelman, M. B., Parks, M. R., and Albrecht, T. L. 1987. Beyond close relationships: Support in weak ties. In T. L. Albrecht and M. B. Adelman (Eds.), *Communicating Social Support.* Newbury Park, CA: Sage. pp. 126–147.

Albrecht, T. L., Burleson, B. R., and Goldsmith, D. 1994. Supportive communication. In M. L. Knapp and G. R. Miller (Eds.), *Handbook of Interpersonal Communication* (2nd ed., pp. 419–449). Thousand Oaks, CA: Sage.

Aviv, R., Erlich, Z., and Ravid, G. 2003. Network analysis of cooperative learning. In *Proceedings of the Fourth ICICTE 2003*, July 3–5, 2003, Samos Island, Greece.

Barnes, M. K., and Duck, S. 1994. Everyday communicative contexts for social support. In B. Burleson, T. Albrecht, and I. G. Sarason (Eds.), *Communication of Social Support: Messages, Interactions, Relationships and Community.* Thousand Oaks, CA: Sage. pp. 175–194.

Becker-Beck, U., Wintermantel, M., and Borg, A. 2005. Principles of regulating interaction in teams practicing face-to-face communication versus teams practicing computer-mediated communication. *Small Group Research*, 36, pp. 499–536.

Bowker, N., and Tuffin, K. 2002. Disability discourses for online identities. *Disability and Society*, 17(3), pp. 327–344.

Bowker, N., and Tuffin, K. 2003. Dicing with deception: People with disabilities' strategies for managing safety and identity. *Journal of Computer-Mediated Communication* 8(2). Retrieved November, 9, 2008 from http://jcmc.indiana.edu/vol8/issue2/bowker.html.

Bradley, N., and Poppen, W. 2003. Assistive technology, computers and Internet may decrease sense of isolation for homebound elderly and disabled persons. *Technology and Disability*, 15(1), pp. 19–25.

Braithwaite, D. O., Waldron, V. R., and Finn, J. 1999. Communication of social support in computer-mediated groups for people with disabilities. *Health Communication*, 11, pp. 123–151.

Brennan, P. F., Moore, S. M., and Smyth, K. A. 1991. ComputerLink: Electronic support for the home caregiver. *Advances in Nursing Science*, 13(4), pp. 14–27.

Buchanan, H., and Coulson, N. S. 2007. Accessing dental anxiety online support groups: An exploratory qualitative study of motives and experiences. *Patient Education and Counseling*, 66, pp. 263–269.

Cobb, S. 1976. Social support as a moderator for life stress. *Psychosomatic Medicine*, 38, pp. 300–314.

Coulson, N. S. 2005. Receiving social support online: An analysis of a computer-mediated support group for people with irritable bowel syndrome. *CyberPsychology and Behavior*, 10, pp. 147–150.

Coulson, N. S., Buchanan, H., Aubeeluck, A. 2007. Social support in cyberspace: A content analysis of communication within a Huntington's disease online support group. *Patient Education and Counseling*, 68, pp. 173–178.

Craven, P., and Wellman, B. 1973. The network city. *Sociological Inquiry*, 43(3), pp. 57–88.

Cutrona, C. E., and Suhr, J. A. 1992. Controllability of stressful events and satisfaction with spouse support behaviors. *Communication Research*, 19, pp. 154–174.

Eastin, M. S., and LaRose, R. 2005. Alt.support: Modeling social support online. *Computers in Human Behavior*, 21, pp. 977–992.

Ellison, N. B., Steinfield, C., and Lampe, C. 2007. The benefits of Facebook "friends:" Social capital and college students' use of online social network sites. *Journal of Computer-Mediated Communication*, 12(4), article 1. Retrieved November 9, 2008 from http://jcmc.indiana.edu/vol12/issue4/ellison.html.

Fisher, D., Smith, M., and Welser, H. T. 2006. You are who you talk to: Detecting roles in Usenet newsgroups. In *Proceedings of the 39th Hawaii International Conference on System Sciences*. Kauai, Hawaii, Jaunary 4–7. Los Alamitos, CA: IEEE Press.

Ford, L. A., Babrow, A. S., and Stohl, C. 1996. Social support messages and the management of uncertainty in the experience of breast cancer: An application of problematic integration theory. *Communication Monographs*, 63, pp. 189–207.

Fox, S. 2004. *Older Americans and the Internet*. Washington, DC: Pew Internet and American Life Project.

Garton, L., Haythornthwaite, C., and Wellman, B. 1997. Studying online social networks. *Journal of computer-mediated communication*, 3(1). Retrieved November 9, 2008 from http://jcmc.indiana.edu/vol3/issue1/garton.html.

Goldsmith, D. 1992. Managing conflicting goals in supportive interaction: An integrative theoretical framework. *Communication Research*, 19, pp. 264–286.

Greenfield, P., and Subrahmanyam, K. 2003. Online discourse in a teen chatroom: New codes and new modes of coherence in a visual medium. *Journal of Applied Developmental Psychology*, 24(6), pp. 713–738.

Hampton, K. N., and Wellman, B. 2000. Examining community in the digital neighborhood: Early results from Canada wired suburb. In T. Ishida and K. Isbister (Eds.), *Digital Cities: Technologies, Experiences, and Future Perspectives*. Heidelberg, Germany: Springer-Verlag. pp. 194–208.

Harris, S. 2006. Emotional support on the Internet V1.23. Retrieved November 9, 2008 from http://www.compulink.co.uk/~net-services/care/list.htm.

Haythornthwaite, C. 2000. Online personal networks: Size, composition and media use among distance learners. *New Media and Society*, 2(2), pp. 195–226.

Heitzmann, C. A., and Kaplan, R. M. 1988. Assessment of methods for measuring social support. *Health Psychology*, 7(1), pp. 75–109.

Ickes, W. (Ed.) 1997. *Empathic Accuracy*. New York: The Guilford Press.

Jeong, A. 2005. Discussion analysis tool (DAT). Retrieved November 9, 2008 from http://garnet.fsu.edu/~ajeong.

Jones, Q., Ravid, G., and Rafaeli, S. 2004. Information overload and the message dynamics of online interaction spaces: A theoretical model and empirical exploration. *Information Systems Research*, 15(2), pp. 194–210.

Joyce, E., and Kraut, R. E. 2006. Predicting continued participation in newsgroups. *Journal of Computer-Mediated Communication*, 11(3), article 3. Retrieved November 9, 2008 from http://jcmc.indiana.edu/vol11/issue3/joyce.html.

Kalman, Y. M., Ravid, G., Raban, D. R., and Rafaeli, S. 2006. Pauses and response latencies: A chronemic analysis of asynchronous CMC. *Journal of Computer-Mediated Communication*, 12(1), article 1. Retrieved November 9, 2008 from http://jcmc.indiana. edu/vol12/issue1/kalman.html.

Kanayama, T. 2003. Ethnographic research on the experience of Japanese elderly people online. *New Media and Society*, 5(2), pp. 267–288.

Kaplan, B. H., Cassel, J., and Gore, S. 1977. Social support and health. *Medical Care*, 15, pp. 47–58.

Kavanaugh, A., and Patterson, S. 2001. The impact of community computer networking on community involvement and social capital. *American Behavioral Scientist*, 45, pp. 496–509.

Kiesler, S., and Kraut, R. 1999. Internet use and ties that bind. *American Psychologist*, 54, pp. 783–784.

Klemm, P., and Wheeler, E. 2005. Cancer caregivers online: Hope, emotional roller coaster, and physical/emotional/psychological responses. *Computers, Informatics, Nursing: CIN* 23(1): pp. 38–45.

Kraut, R., Lundmark, V., Patterson, M., Kiesler, S., Mukopadhyay, T., and Scherlis, W. 1998. Internet paradox: A social technology that reduces social involvement and psychological well-being? *American Psychologist*, 53, pp. 1017–1031.

Krippendorff, K. 1980. *Content Analysis: An Introduction to Its Methodology*. Newbury Park, CA: Sage.

La Gaipa, J. J. 1990. The negative effects of informal social support systems. In S. Duck and R. C. Silver (Eds.), *Personal Relationships and Social Support*. London: Sage. pp. 122–139.

Laghos, A., Zaphiris, P. 2006. Sociology of student-centred e-Learning communities: A network analysis. *In Proceedings of the IADIS International Conference*, e-Society 2006, July 13–16, 2006, Dublin, Ireland.

Lakey, B., and Cohen, S. 2000. Social support theory and measurement. In S. Cohen, L. Underwood Gordon, L. G. Underwood, and B. H. Gottlieb (Eds.), *Social Support Measurement and Intervention* Oxford University Press, pp. 29–52.

Lange, P. G. 2007. Publicly private and privately public: Social networking on YouTube. *Journal of Computer-Mediated Communication*, 13(1), article 18. Retrieved November 9, 2008 from http://jcmc.indiana.edu/vol13/issue1/lange.html.

Lapadat, J. 2002. Written interaction: A key component in online learning. *Journal of Computer-Mediated Communication*, 7(4). Retrieved November 9, 2008 from http:// jcmc.indiana.edu/vol7/issue4/lapadat.html.

Leatham, G., and Duck, S. 1990. Conversations with friends and the dynamics of social support. In S. Duck (Ed.), *Personal Relationships and Social Support*. London: Sage. pp. 1–29.

Maloney-Krichmar, D., and Preece, J. 2005. A multilevel analysis of sociability, usability, and community dynamics in an online health community. *Transactions on Computer-Human Interactions* (TOCHI), 12(2), pp. 201–232.

Mayring, P. 2000. Qualitative content analysis. *Forum: Qualitative Social Research*, 1(2). Retrieved November 9, 2008 from http://www.qualitative-research.net/fqs-texte/2-00/ 2-00mayring-e.htm.

McMellon, C. A., and Schiffman, L. G. 2002. Cybersenior empowerment: How some older individuals are taking control of their lives. *Journal of Applied Gerontology*, 21(2), pp. 157–175.

Mickelson, K. 1997. Seeking Social Support: Pareals in electronic support groups. *In Culture of the Internet*, S. Kiesler (Ed.) Mahwah, NJ: Lawrence Erlbawn Associates, pp. 157–178.

Moursund, J. 1997. SANCTUARY: Social support on the Internet. In J. E. Behar (Ed.), *Mapping Cyberspace: Social Research on the Electronic Frontier* (pp. 53–78). Oakdale, NY: Dowling College.

Nahm, E. S, Resnick, B., and Mills, M. E. 2003. A model of computer-mediated social support among older adults. In *Annual Symposium Proceedings of the American Medical Informatics Association*, p. 948.

Nie, N. H. 2001. Stability, interpersonal relationships and the Internet: Reconciling conflicting findings. *American Behavioral Scientist*, 45, pp. 420–435.

Office of Communication. 2006. *Consumers and the Communications Market: 2006*. Ofcom Consumer Panel. Retrieved August 31, 2009 from http://www.ofcom.org.uc/research/cm/cm06/cmo6-print/man.pdf.

O'Neill, J., and Martin, D. 2003. Text chat in action. In Proceedings of the International ACM SIGGROUP Conference on Supporting Group Work. Sanibel Island, Florida, November 9–12, 2006. pp. 40–49.

Pfeil, U., and Zaphiris, P. Patterns of empathy in online communication. In Proceedings of CHI 2007—the ACM Conference on Human Factors in Computing Systems, April 28, 2007 through May 3, 2007. San Jose, CA, CHI '07 (pp. 919–928). New York: ACM Press.

Pfeil, U., Zaphiris, P., and Wilson, S. 2009. Older people's perceptions and experiences of online social support. *Interacting with Computers*. Retrieved February 24, 2009 from http://dx.doi.org/10.1016/j.intcom.2008.12.001.

Preece, J. 1998. Empathetic communities: Reaching out across the Web. *Interactions*, 2, pp. 32–43.

Preece, J. 1999. Empathic communities: Balancing emotional and factual communication, *Interacting with computers*, 12(1), pp. 63–77.

Preece, J. 2000. Empathy online. *Virtual Reality*, 4, pp. 74–84.

Preece, J., and Ghozati, K. 1998. Offering support and sharing information: A study of empathy in a bulletin board community. In Conference of Collaborative Virtual Environments, Manchester, England.

Preece, J., and Ghozati, K. 2001. Observations and explorations of empathy online. In. R. R. Rice and J. E. Katz (Eds.), *The Internet and Health Communication: Experience and Expectations* (pp. 237–260). Thousand Oaks, CA: Sage.

Rafaeli, S., and Sudweeks, F. 1997. Interactivity on the net. In F. Sudweeks, M. McLaughlin, and S. Rafaeli (Eds.), *Network and Netplay: Virtual Groups on the Internet*. Menlo Parks, CA: AAAI/MIT Press.

Ridings, C. M., and Gefen, D. 2004. Virtual community attraction: Why people hang out online. *Journal of Computer-Mediated Communication*, 10(1). Retrieved November 9, 2008 from http://jcmc.indiana.edu/vol10/issue1/ridings_gefen.html.

Robinson, D. 1988. Self-help groups. In R. S. Cathcart and L. A. Samovar (Eds.), *Small Group Communication: A Reader* (pp. 117–129). Dubuque, IA: Wm. C. Brown.

Rodgers, S., and Chen, Q. 2005. Internet community group participation: Psychosocial benefits for women with breast cancer. *Journal of Computer Mediated Communication*, 10, p. 4. Retrieved November 9, 2008 from http://jcmc.indiana.edu/vol10/issue4/rodgers.html.

Sandelowski, M. 2000. Focus on research methods: Whatever happened to qualitative description? *Research in Nursing and Health*, 23(4), pp. 334–340.

Scott, J. 2000. *Social Network Analysis: A Handbook*. (2nd ed.). Thousand Oaks, CA: Sage.

Valkenburg, P. M., and Peter, J. 2007. Online communication and adolescent well-being: Testing the stimulation versus the displacement hypothesis. *Journal of Computer-Mediated Communication*, 12(4), article 2. Retrieved November 9, 2008 from http://jcmc.indiana.edu/vol12/issue4/valkenburg.html.

Wallace, P. 1999. *The Psychology of the Internet*. Cambridge, MA: Cambridge University Press.

Walther, J. B. 1996. Computer-mediated communication: Impersonal, interpersonal, and hyperpersonal interaction. *Communication Research*, 23, pp. 3–43.

Walther, J. B., and Boyd, S. 2002. Attraction to computer-mediated social support. In C. A. Lin and D. Atkin (Eds.), *Communication Technology and Society: Audience Adoption and Use*. Cresskill, NJ: Hampton Press. pp. 117–129.

Wellman, B., and Gulia, M. 1999. Net surfers don't ride alone: Virtual communities as communities. In M. A. Smith and P. Kollock (Eds.), *Communities in cyberspace*. London: Routledge. pp. 167–194.

White, M. H., and Dorman, S. M. 2000. Online support for caregivers: Analysis of an Internet Alzheimer mailgroup. *Computers in Nursing*, 18, pp. 168–176.

Winefield, H. R. 2006. Support provision and emotional work in an Internet support group for cancer patients. *Patient Education and Counseling* 62, pp. 193–197.

Winefield, H. R., Coventry, B. J., Lewis, M., Harvey, E. J. 2003. Attitudes of breast cancer patients to support groups. *Journal of Psychosocial Oncology* 21, pp. 39–54.

Wright, K. B. 1999. Computer-mediated support groups: An examination of relationships among social support, perceived stress, and coping strategies. *Communication Quarterly*, 47(4), pp. 402–414.

Wright, K. B. 2000a. The communication of social support within an on-line community for older adults: A qualitative analysis of the SeniorNet community. *Qualitative Research Reports in Communication*, 1(2), pp. 33–43.

Wright, K. B. 2000b. Computer-mediated social support, older adults, and coping. *Journal of Communication*, 50(3), pp. 100–118.

Xie, B. 2005. Getting older adults online: The experiences of SeniorNet (USA) and OldKids (China). In B. Jaeger (Ed.), *Young Technologies in Old Hands—An International View on Senior Citizen's Utilization of ICT*. Copenhagen, Denmark: DJOF Publishing. pp. 175–204.

Xie, B. 2007. Older Chinese, the Internet, and well-being. *Care Management Journals: Journal of Long Term Home Health Care*, 8(1), pp. 33–38.

Xie, B. 2008. Multimodal computer-mediated communication and social support among older Chinese Internet users. *Journal of Computer-Mediated Communication*, 13, 728–750.

Zaphiris, P., and Sarwar, R. 2006. Trends, similarities and differences in the usage of teen and senior public online newsgroups. *ACM Transactions on Computer-Human Interaction (TOCHI)*, 13(3), pp. 403–422.

Chapter 7

Online Content Sharing

Masahiro Hamasaki, Kouchiro Eto, Sri Kurniawan, Tom Hope, Hideaki Takeda, and Takuichi Nishimura

Contents

7.1 Introduction

In recent years, we have seen the rapid emergence of user generated content (UGC), a philosophy that radically alters how information is created and shared on the Internet. UGC denotes content that is created by ordinary people instead of professional artists, writers, or journalists. The World Wide Web, and especially Web 2.0, enables massive interactive collaboration through large-scale bulletin boards and social tagging, which is impossible in most other communication channels. It also enables large-scale sharing of information beyond simple text (e.g., picture sharing in Flickr™, video sharing in YouTube™, and three-dimensional [3D] model sharing in *Second Life*—although the idea of content sharing in *Second Life* has shifted into more explicit "content trading.") New tools and sites that facilitate collaboration, publication, and distribution of content by and among individuals have been widely promoted, culminating in the acquisition of YouTube™ by Google™ for $1.65 billion in 2006*. Other systems include Sodaplay (Burton et al. 2008) and Springs World 3D (Falco 2008) by the Soda company, which enable users to create a model and share it via the Internet. Waddoups has developed Juice (Waddoups 2008), with which users produce a model and move it under physical law constraints. In 1999, SCEI Corporation began selling Panekit (SCEI 1999), a toolkit to create vehicles of various kinds such as cars, ships, and airplanes by combining square panels.

The chapter is organized as follows. We begin by defining online content sharing and exploring some of the intellectual property (IP) issues. In Section 7.3, two case studies are presented that highlight the problems and potential technological solutions of online content sharing and copyright. The first introduces the Japanese video sharing Web site, Nico Nico Douga (NND), which we use as an illustration of how IP is currently recognized in the online content sharing world. In this site users reuse others' video clips to create new videos, with two different methods of crediting the original author. In the second case study, we present another method of approaching the problem of creation and copyright, by describing and analyzing a content-sharing platform that we developed, called Modulobe. This consists of a 3D modeling tool and a model-sharing Web site in which author crediting is enforced automatically in the source system. Finally, in Section 7.5, we summarize the issues raised.

7.2 UGC and Creation

Online content sharing arguably started in 1945 when Vannevar Bush described Memory Extender (Memex; Bush 1945), an electromechanical device that an individual could use to read a large self-contained research library and add or follow associative trails of links and notes created by that individual or recorded by other

---

* http://www.google.com/press/pressrel/google_youtube.html

researchers. Ted Nelson coined the term "transclusion" as well as "hypertext" and "hypermedia" (Nelson 1994), and Tim Berners-Lee proposed the World Wide Web in 1989. The World Wide Web began in 1991 as an information-sharing service on the Internet, but the early Web could share content only in the form of text. In 1992, Marc Andreessen developed a Web browser that could handle images in addition to text, making it possible to engage in online content-sharing activities that are quite close to what we do in recent years.

Through the explosive growth of the Internet in the mid-1990s, people have created text and various media and published them in multiple places such as personal blogs, Wikipedia,* Digg,† and other free content-sharing Web sites. Furthermore, advanced search engines, social bookmarking, and social network services have accelerated their distribution. Such creative activity by the general public is known variously as UGC, UCC (user created content), and CGM (consumer generated media) and is gaining much attention from the media and developers.

The OECD (2007) defines three central characteristics for UGC as follows:

Publication requirement: While theoretically UCC could be made by a user and never actually be *published* online or elsewhere, we focus here on the work that is published in some context, be it on a publicly accessible Web site or on a page on a social networking site only accessible to a select group of people (i.e., fellow university students.) This is a useful way to exclude email, bilateral instant messages, and the like.

Creative effort: This implies that a certain amount of creative effort was put into creating the work or adapting existing works to construct a new one; i.e., users must add their own value to the work. The creative effort behind UCC often also has a collaborative element to it, as is the case with Web sites that users can edit collaboratively. For example, merely copying a portion of a television show and posting it to an online video Web site (an activity frequently seen on the UCC sites) would not be considered UCC. If a user uploads his or her photographs, however, expresses his or her thoughts in a blog, or creates a new music video, this could be considered UCC. Yet the minimum amount of creative effort is hard to define and depends on the context.

Creation outside professional routines and practices: User-created content is generally created outside professional routines and practices. It often does not have an institutional or a commercial market context. In the extreme, UCC may be produced by nonprofessionals without the expectation of profit or remuneration. Motivating factors include connecting with peers, achieving a certain level of fame, notoriety, or prestige, and the desire to express oneself.

---

* http://www.wikipedia.org/
† http://digg.com/

The Web has become a common platform for communication and presents various advantages over more traditional methods. For example, it erases the sense of distance between people because, provided access is available, people in any location can communicate with each other. It enables massive interactive collaboration through large-scale bulletin board systems (BBSs) and social tagging, which is impossible through real-world communication channels. Additionally, it affords large-scale sharing of information beyond text, such as multimedia content.

Such new features of communication have fostered new styles of creative activity. Within this, a significant feature is the massively collaborative creation of digital content. People now create new content by communicating or collaborating with many others via the Web. A new aspect of this collaboration is the style of participation and its effects. For example, because vastly numerous people are involved, they often do not know each other at all. In terms of the content itself, many evolved or inspired versions can be created, such as when a video explores a new idea or catches on. Another difference with older methods of communication is the digital reuse of content. Some components of created work are reused in subsequent evolved or inspired versions. For example, an image or sound in a video can be reused in a new version of the video. Consequently, digital work can be developed collaboratively via the Web in ways that are not possible offline.

7.2.1 Types of UGC

The popularity of UGC had led to the rise of different types of content being shared online. We list some of the most popular types of UGC and the sites that most people are familiar with. Some of these UGC are discussed in other chapters of this book.

7.2.1.1 Discussion Boards

Sometimes referred to as "discussion groups," "discussion forums," "message boards," and "online forums," discussion boards are asynchronous communication systems that allow a member of that forum to post a comment, idea, or question online. Other members read that posting and respond with their own remarks and ideas over time. Discussion boards carry historical value because they provide the context for the original digital communities. Arguably, discussion boards started in mid-1970s as the virtual bulletin boards known as the Usenet, in which anyone with access to the network can post a message to any "newsgroup" at any time (Hartman and Koohang 2005).

7.2.1.2 Blogs

Blogs, or weblogs, are frequently updated Web pages written by ordinary people (as opposed to professional writers, although many companies nowadays maintain a blog site) with a series of archived posts, typically in reverse chronological order. Most posts are textual, though people increasingly include other multimedia content, and most contain a strong sense of the author's personality, passions, and point of view (Nardi et al. 2004). Blogs started around 1997, with Dave Winer's Scripting News (still the longest running blog), which includes his reflections and commentaries on a wide range of topics (Schiano et al. 2004).

7.2.1.3 Wikis

Made popular by Wikipedia, the world's largest collaboratively edited source of encyclopedic knowledge (Völkel et al. 2004), wikis are simple content management systems geared toward enabling readers to modify the content of the Web site easily. Wikis were first introduced by Ward Cunningham in 1994 within the programming language patterns group (Cunningham and Leuf 2001).

7.2.1.4 Social Network Sites

Social network sites (SNS) are Web-based services that allow individuals to (1) construct a public or semipublic profile, (2) articulate a list of users to share a connection with, and (3) view and traverse their list of connections and those made by others within the system (Boyd and Ellison 2007). MySpace™ and Facebook™ are, at the time of writing, the two SNS with the highest number of members, although other sites such as Orkut or LinkedIn™ have also attracted a large number of members. The first recognizable SNS, Sixdegrees.com (the site is no longer active), launched in 1997. There are two types of SNSs: those adopting open registration (enabling strangers to connect with each other through common interests) and those that are invitation only (providing maintenance to preexisting real-life social networks; Wan et al. 2008). As the number of sites has grown, some SNSs decidedly have adopted different focuses, for example, LinkedIn, which is based on developing business and employment networks. Regardless of the starting point, SNSs normally evolve from having a shared background or interest into social relationships and connections with people. Hence, online communication slowly changes from being merely task based or for sharing information into an end in itself.

7.2.1.5 Domain-/Product-Specific Sharing Sites

Increasingly, people share experiences, pictures, video clips and other creations online. Examples include Google's Picasa™, Flickr™ (arguably the most popular photo-sharing site), and tripadvisor (where users share their experiences with certain hotels, restaurants, tourist attractions, etc.)

7.2.2 Privacy, IP, and Sharing Online Content

UGC has generally promoted the idea of the "wisdom of crowds" (Surowiecki 2004) or collective intelligence, as it allow experts in a certain field or topic to contribute content for others to consume and improve upon; but UGC additionally presents new challenges around issues such as IP,* privacy, and defamation, where the ethical and legal rules are not yet clearly defined. These problems can occur, as the nature of the interactions (between the creators and the consumers of information) and information creation and consumption itself are very different to those of more traditional online systems. For example, the recent proliferation of digital cameras, and in particular camera phones, has made it possible to almost instantly upload and share photographs of events in one's personal life, without the need to take time for developing, or in many cases, editing, the images. This has in many cases meant that photos, which would in the past have remained private, are publicly shared. No longer are photographs limited to "Kodak moments" shared with friends (Chaflen 1987); now, publicly distributed photographic content includes many images of formerly private spaces and behavior (Miller 2007). This mixing of public and private can have consequences for both the creators of content and the online technologies that enable their sharing. In some cases, with the widening collection of contextual data (such as location through the global positioning system [GPS] on mobile phones), sharing of content produces privacy issues that users find worrying (Ahern et al. 2007).

In addition to concerns over privacy in the viewing of uploaded videos and photos, the ease with which viewers of digital content can download, save, and modify the files they find online brings forth serious problems of IP. Sites such as YouTube, as they have become popular, have needed to concede to the demands of large producers and distributors of films and music, who, due to their financial and legal strength, persuade the Web sites to remove copyrighted content. In cases like this, there are clear legal avenues that the companies can take to demonstrate their ownership of the digital content and rights to distribute it to consumers as they wish. However, amateur filmmakers, writers, photographers, and producers of other digital work often have little recourse other than to contact the Web sites and ask to be credited for the work or take potentially expensive legal action. As Internet technology is still relatively "new," IP issues form an ever-changing labyrinth that can be difficult for everyday users to navigate their way through. Indeed, the issues of UGC and copyright infringement have not yet been resolved internationally, let alone on national levels (Samuelson 1995; Osbourne 2008); so it remains to be seen what protection traditional approaches will offer to users and the companies that host UGC who are expressing anxiety about the legal implications (Holmes and Ganley 2007).

There are two primary approaches taken online to issuing online content and protection of copyright. Both of these can be seen to be extensions of politics and legal approaches to creators' rights and IP that existed predominantly offline and as such

---

* Strictly speaking, in this chapter we are discussing content sharing and copyright issues, but in some cases (e.g., science papers, online education, and business-oriented blogs), it may be more correct to use the broader definition of intellectual property.

are still being debated in relation to online environments (Chadwick 2006). In many of these debates, the central issue revolves around copyright, fair use, and derivative works. These cases are important to resolve, as in some cases, for example, online educational courses, a number of works may be used and combined to form new content, and bulletin boards may contain thousands of messages created by students (Harrison and Stephen 1996). The fact that content can be modified so easily, makes earlier IP legal frameworks difficult to apply, and responses to the challenge of protecting users' rights while providing freedom to create have primarily been met by enforcing the law either through the signing of agreements or via technological measures.

The first of these approaches that are most clearly directed toward digital content is the recent development of Creative Commons (CC)* licensing agreements, an approach that aims to allow flexibility and ease of use for nonprofessional and professional creators in online environments. CC, a nonprofit organization founded in 2001, provides a range of licenses (i.e., CC licenses), which can be selected by users to fit their wishes over the use of their created work. The organization claims that the licenses do not replace traditional copyright, but "work alongside copyright, so you can modify your copyright terms to best suit your needs."† The licenses are designed to be acceptable globally, thus protecting user rights within a global legal framework.

In addition to licensing approaches, such as CC, more obviously technological methods are used to protect rights over shared content online. The most prominent—and contentious—of these is digital rights management (DRM) encryption. This technology is primarily an attempt to restrain users from unauthorized reproduction of music and video, though recently it is also applied to e-books. DRM can also be used to limit the time that content can be viewed and listened to and the systems that it can be accessed on. This brings to the fore concerns by many that the control over online content is far stricter than it was for offline content, with technological means being used in place of legal means (Chadwick 2006). CC can be seen in part as a reaction to DRM, which is often static in nature—i.e., once DRM is applied to a digital file, it is difficult to remove it. Nevertheless, DRM is a valid technique to prevent piracy and protect the rights (moral and financial) of creators and undoubtedly is here to stay for the near future.

These two approaches to copyright can tend to fall down when the works or parts of them are used to create new content, such as in video mash ups, sampling of audio tracks, or using parts of images—digitally modified—to generate new pieces. However, this interactive relationship with content (often simply essays in a blog or sampling sections of music to create new songs) is at the heart of the philosophy of Web 2.0 and the

---

* http://creativecommons.org
† ibid. CC is only one recent attempt at formulating new licensing, but it is one of the most widely used, having an estimate 130 million licensed works in 2008. In 2005, the organization was involved in setting up Science Commons (http://sciencecommons.org), which aims to do the same for scientific research as it is doing for other creative work.

new way Web sites are designed to be used (Oreilly 2007). There is a clear need therefore to understand how new content sharing occurs and what measures can be taken to ensure protection of creators' rights while promoting sharing and innovation.

As early as thirty years ago, some initiatives were proposed to deal with the IP rights of online content. Ted Nelson introduced a system named "Xanadu" (Nelson 1994), a platform to manage shared content and their IPs. One of the most important features of Xanadu is "transclusion." For example, an article might include a picture or a paragraph from a different article. In ordinary circumstances, the included data are copied and stored in two places. However, transclusion allows it to be stored only once and viewed in different contexts using a modular design. With this, Nelson suggested that micropayments could be automatically exacted from the reader for all the text, no matter how many parts of the content are taken from various places. This type of modular design has yet to be realized, though potential can be seen in sites that either function as aggregators or have this as part of their design.*

7.3 Content Sharing and Social Networks

7.3.1 The Place of Social Networking in Online Content

UGC is at heart a collaborative enterprise. Although content can be created by an individual, the intention is often to reach a select audience, leading to interaction, which may subsequently inspire further content. Photograph-sharing Web sites such as Flickr and Fotolog provide the space for users to upload their own photographs and give opportunities for interaction through commenting systems. Many of these sites operate a two-tier pricing system, giving users the chance to have a free account of limited storage or more space with regular payments.

Some studies have highlighted that there are recipes for successful UGC sites. A paper by Silva and Dix (2007) summarized that the success of YouTube (the most visited Web site excluding search engines) is due to three factors:

A new type of Web user: those who count Web surfing among their hobbies and who, in some way, socialize via the Web.
Content: ranging from the personal broadcasting aspect to sharing movies that they love with the world
Fitness for purpose: which is to explore, have fun, and enjoy the "route" as much as the goal.

---

* At the time of writing, this can be seen in two ways: sites that collect data from other places on the Internet and display them together on one page, and sites that aim to become modules in other Web pages. Some social networking sites and search engines (e.g., www.facebook.com, www.google.com) are doing both, with the latter taking the form of common identification and password login infrastructure, which allows the subsequent collection of users' searches and favorite content.

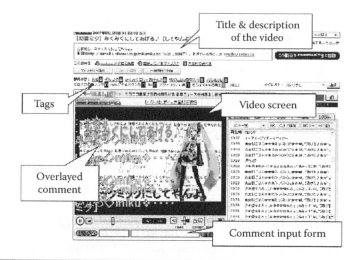

Figure 7.1   Nico Nico Douga video-sharing Web site.

7.3.2 Case Study: NND

In this section we explore NND,* a Japanese video-sharing Web site, and an example of a recent method of dealing with the collaborative creation and sharing of content that attempts to (at least partially) solve the problem of derivative works. In fact, the principal feature of the Web site is users are able to, and regularly do, reuse others' content to create their own.

NND can be seen as a Japanese YouTube. It became the most popular video-sharing Web site in Japan after it was founded in December 2006. By July 1, 2008, it had more than 7.9 million users and had published more than 0.8 million videos (Nakamura et al. 2008). At the time of writing, one function that differentiates NND from YouTube is the function to overlay comments directly onto videos. Users can add comments about an image, sound, or scene in the video at the specific playback time where it occurs (see Figure 7.1). The creator of the video and other viewers can interact with these by adding more comments.† These comments take the forms of questions, jokes, imitating words on the video, or emoticons representing different reactions. The comments themselves can in turn inspire the original creator or other creators to produce new videos in response.

---

* http://www.nicovideo.jp/ The name of the site means "smiling videos" or "videos that make you smile."

† One often repeated concern of researchers has been the difficulty of creating a sense of awareness and "presence" with asynchronous communication in online environments (Dourish 2006). In allowing other users to leave comments at the actual points of interest, Nico Nico Douga may be said to be partially solving this issue.

On the site, one of the most popular types of video is the so-called MAD movie. These are fan-made videos, produced by combining video clips and sounds taken from *anime* (animation) videos to create a new clip that is different from the original. This kind of content is often referred to as a mash up style of video creation, i.e., a video is created by mashing up existing videos, sounds, and images. Informal interviews with users indicate that the creators of MAD movies are inspired by one another's works and many are repeat contributors.* Although MAD movies are attractive to many viewers due to their use of extracts from popular commercial anime programs (as viewers can see something familiar that has been creatively altered), these movies are problematic from the point of view of IP, and some creators have received harsh criticism from the commercial companies that produce the original anime programs.

7.3.2.1 The Hatsune Miku Phenomenon and NND

In August 2007, the Yamaha Corporation released Vocoloid 2 software, a singing synthesizer engine. Crypton Future Media Inc. developed an instalment of Vocoloid 2 called Hatsune Miku, which means "first sound of the future." It features the sampled voice of the Saki Fujita, which forms the data for creating synthesized singing. The software is designed to be used to create songs and is particularly suited to the types of pop songs found on the soundtracks to popular anime videos. Users can enter lyrics, and a melody and a realistic-sounding singing voice subsequently performs them. It has proved to be very successful, and much of the success is due to the NND video-sharing Web site.

Soon after the release of the Hatsune Miku, users of NND began posting videos with songs created with it. NND became a central place for people to share their Hatsune Miku videos and songs, which spurned responses from other users, who would remix them, add illustrations, and upload new videos, asking for comments. Even though Hatsune Miku's original purpose was for song sharing, soon people were sharing many different types of content. For example, initially, Vocaloid only had one mascot character, which was printed on the software package. However, soon people proposed different images, followed by music video clips, 3D animations, and illustrations to accompany songs. Different types of creators also collaborated to improve the Hatsune Miku character (an illustrated girl with green hair): songwriters from the computer music field, illustrators who come from "doujinshi"—self-published manga culture—and computer graphics (CG) creators. This content continued to be created predominantly by amateurs, although some professionals are known to have contributed to the Hatsune Miku phenomenon. A detailed description of Hatsune Miku, including the data analysis of NND (using social network analysis techniques) to understand how different types of creators interact to create new content through their social network, is published in a study by Hamasaki et al. (2008).

---

* http://www.goldsmiths.ac.uk/media-research-centre/project2.php

7.3.2.2 NND, Copyright, and Sharing

Several important issues arise from the studies that we did on NND and Hatsune Miku. First, through an analysis of following hyperlinks pointing to original creators' content, we established that derivative works form a key part of UGC in this context. Fundamentally, this takes the shape of different communities that interact with each other but often support similar types of content sharing, as was the case in one particular type of creation community that focused on creating illustrations. The sheer number of related videos and songs may be unusual—indeed Hatsune Miku is a movement of some sorts—but with the potential for created work to be bought and sold, the importance of recognizing creators' rights is obvious. NND has been forced to take steps to remove copyrighted content belonging to the major commercial distributors, partly due to pressure from YouTube,* but it is the ordinary users who may suffer should their content not be properly licensed structurally.

In a move to tackle the copyright problem, the Web site has recently introduced Niconi Commons,† where users can upload videos specifically with the intention of others using them (under a CC license.) Each video has a unique identification (ID), and so, if other users register their use of a particular video, it is possible to visualize the "tree" of derivative works. This visualization, the licensing, and use of its features is entirely up to the discretion of NND users. As such, although it can be useful to see connections between some shared content, it is not possible to say that the whole NND system supports the rights of creators. Other than visualization, then, it does not substantially change the state of affairs. It is clear that the site is successful for many different reasons, but as yet, relations between copyright and sharing of content created by different communities are not fully supported.

7.3.3 Case Study: Modulobe

Another content-sharing system that allows users to share 3D models is the system that we developed, called Modulobe. Modulobe is based on the idea that instead of focusing on the appearance of a 3D model, we focus on the movements of the "objects" created. The basic idea is that one can recognize the object and its structure through the motion of bright points on a black background (Johansson 1973). In real life, for example, such an image sequence can be obtained using motion capture systems, which capture only specific small reflective cubes on the subjects' body. The cubes are usually put on the ankle or waist where the angles of the body parts change. People cannot estimate the structure by looking at a static scene, but when the object moves, most people can recognize its structure. Therefore, in Modulobe, "objects" are created only with shafts (of various angles) and links that move following the laws of physics.

---

* http://d.hatena.ne.jp/metagold/20080705
† http://www.niconicommons.jp/

Although it would be simpler to use existing tools, some of these tools cannot cope with the computation costs involved with simulating movements. For example, it is complex for the systems developed by Burton et al. (2008) and Falco (2008) to present rigid bodies because they use a penalty method for computation of physical simulation. Panekit (SCEI 1999) limits the variety of parts: A user cannot develop new parts. Juice uses open dynamics engine (ODE; http://ode.org/), which uses analytical methods. In this case, the computation effort increases greatly when the number of parts increases. Consequently, a user must use a small number of parts. Therefore, we decided to develop Modulobe* and launched the site for the creation and sharing of models in 2006.

Modulobe has two components: a 3D modeling application to simulate complex motions and a model-sharing Web site to upload and download models complemented with tags, comments, and popularity ratings designed to stimulate users to create new models. The simulation engine is designed to reduce computation costs while still producing realistic-looking models. Since its launch, the Modulobe application has been downloaded 171,267 times, and the Web site has been accessed 686 million times by approximately 100,000 unique internet protocol addresses. More than 3,000 models have been uploaded.

The median number of modules in a model is 31 and the maximum is 1,729. There are 57 models that have over 500 modules. Similarly, the median number of link modules, a special module to produce motion, is 10 and the maximum is 1,662. About 90% (3,019 models) of all models have the link module. In general, creating 3D models with motion is a complex undertaking. However, we noticed that many new users created 3D models with motion and published them to the Web site, which can be an indication of how the functions we implemented are, to a certain extent, helpful for nonprofessional modelers.

7.3.3.1 Modulobe Design

Modulobe comprises a platform for users to create a simple model with complex motions, but it does not particularly address the texture or other fine-detailed properties of the models. We were influenced by the idea of Lego bricks,† in which very simple forms of bricks can result in very complex structures. Particularly inspiring was Papert, who found that Lego blocks can stimulate creativity in children (Wiencek 1987). With Modulobe we would like to extend users' creativity by providing a content-sharing mechanism so that users can build on each other's moving models. The comparable objects to Lego bricks in our system are called "modules."

A solid block like a Lego brick is appropriate for presenting shapes that have volume. However, it is difficult to present a shape that can bend like a hinge. Similarly, a flat panel makes it difficult to present the motion of a hinge. A stick whose cross

---

* http://www.modulobe.com
† http://www.lego.com/

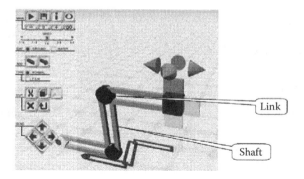

Figure 7.2 Shafts and links.

section is a circle or square makes it difficult to determine the direction of modules. For those reasons, we designed the shape of a module to resemble a slender board.

Modules consist of two types: shafts and links (see Figure 7.2). A link connects shafts with various angles. There are two types of links: static links (to branch a model) and powered links (to set motions.) A user can add four shafts to a static link and two shafts to a powered link (and bend it.) The motion of a link is initiated by setting patterns of an angle of the powered link. There is no limit to the number of shafts and links that a user can use.

The movements of a model built in Modulobe follow the laws of physics: gravitation, inertia, conservation of momentum, and friction with the ground and rebounding. Although the system sounds simple, the simulation engine behind Modulobe required careful design, as it has to function in real time. As an illustration, in a model, the only movable point is a link. Therefore, the motion of a model can be simulated using the penalty method. However, some problems can occur if we use only a simple penalty method for simulation (Baraff 1989). Therefore, we had to modify the method, as described below.

Although a user can create a 3D model, creating this type of model is more complex than creating two-dimensional (2D) models. Therefore, we decided to design an interface in which the default setting is for users to create 2D models (although users can override the default setting if they would like to create a 3D model immediately.)

The second part of Modulobe is the content-sharing section, which is a public site where users can view all of the uploaded models that had been developed using the modeling application. In addition to viewing the downloaded models, users can edit the models and upload new versions.

7.3.3.2 *Implementation of Modeling Application*

Modulobe was implemented in Visual C++ with DirectX 9. It consists mainly of a physical simulation engine and a user interface in which a user edits and views models. In this section, we describe an optimization method of physical simulation for widely distributed computers to cope with the large computation costs for this complex motion model and its user interface, which provides gradual complexity presentation so that users can learn with simpler models before going into more complex models.

7.3.3.2.1 Interface

Modulobe has two modes: edit mode and play mode. In the edit mode, a user can create a model by connecting modules and can set motions of the virtual creature in place by simply specifying a change of the angle of hinges. In the play mode, the user can test their models.

7.3.3.2.2 Edit Mode

Figure 7.3 is a screenshot of the start screen in edit mode. As explained earlier, a model consists of many parts called "modules" of the "shaft" and "link" types. In the edit mode, there is one special shaft, called the "core" at the beginning. A user generates a framework of a model by connecting modules to the core. The model has a tree structure whose source node is the core.

Each module has some guides for adding a new module. When users click one of these guides, a new module is added to the point that is indicated by the guide. Users can change the angle of connection by clicking a connector between modules. The range of the angle of rotational direction is free, but an angle with a neighboring module should be less than ± 45. A range of the angle of folding direction is ± 45. A step angle is ± 45.

In addition, the viewpoint of the starting screen of the edit mode provides the viewpoint of the side of a model (See Figure 7.4). Users will create a 2D shape if they have not bent connectors in the direction of depth. The four tetrahedral objects are controllers for changing the angles.

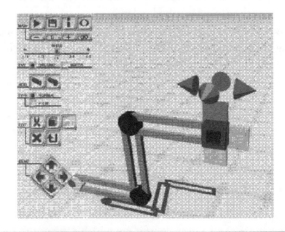

Figure 7.3   Menu for a guide.

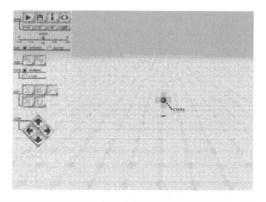

Figure 7.4   Screenshot of the start screen in edit mode. Users can change the mode using the menu on the left.

The center of the link consists only of one rotational direction. The axis of rotation stands vertically on the flat line, which the link module is in. When a user clicks a link, the system shows a graph, which shows a pattern of bending of the link (Figures 7.5 and 7.6). The vertical axis of the graph is time. The horizontal axis of the graph shows the angle of the link. The step angle is ±15, and the maximum angle is ±90. In addition, a user can change the mode, which allows a link to be rotated 360°.

The top and bottom of the timeline are connected. The cursor repeats scrolling from the top to the bottom. The slide bar in the right side is the controller for a cycle of motion of an angle. The default is one cycle per second. The angle that is set up in this controller is the target angle. In the play mode, force is added to modules to be the target angle. Therefore, sometimes a module cannot reach the target angle, especially if the target motion is affected by the gravity.

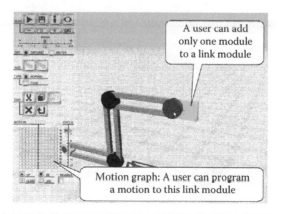

Figure 7.5   Motion graph for programming the motion of a link module.

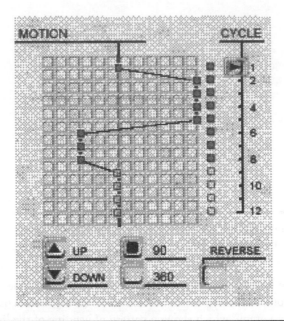

Figure 7.6 Menu for a link. A graph shows the pattern of bending the link.

In addition, a user can make a link that has no target angle. Figure 7.7 portrays a sample of the setting of a pattern of the motion of a hinge. At the second step, the link bends +90 degrees. After four steps, the link bends −45 degrees to the other side. After three steps, the link has no target angle. It becomes free.

A user can reverse this graph, which means that it is easy to create opposite motion on the side of models. For example, the movement of a right leg is generally opposite

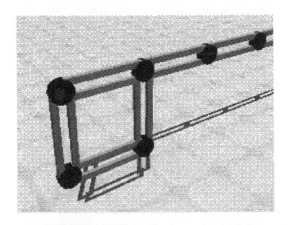

Figure 7.7 Expand mode is off.

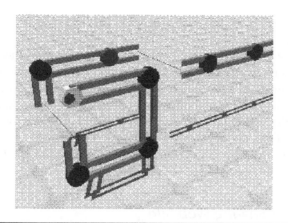

Figure 7.8　Expand mode is on.

that of a left leg when a creature walks. Therefore, in theory, users can easily create a "creature" by creating only half of the "body" and copying and pasting the other half.

It is difficult for a user to control models when some modules are overlapped or crowded (see Figure 7.7). Therefore, Modulobe provides a function to stretch a model to allow users to superimpose modules to create overlap modules as though they were connected (see Figure 7.8).

7.3.3.2.3 Play Mode

In the play mode, users can see animated sequences of the model. The playground consists of a flat ground that has no wall, no obstacles, and no change in topology. In this mode, users can apply an arbitrary force to the models using the drag and drop processes, raise up models by scrolling the mouse wheel (see Figure 7.9), control the speed of operation of models using the speed control slide bar on the upper left of

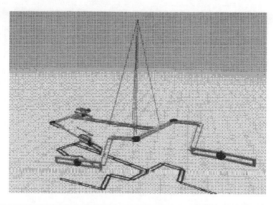

Figure 7.9　Raising up the model.

the screen, and see the relationship among the change of speed and the change of motion of their models.

In this play mode, users can upload a model and an introduction movie that they can produce through the system. Because Modulobe focuses on motions, users can easily make and use a movie as a thumbnail. When the system takes a movie of the model, users can move the model and control the camera's angle. After taking a movie, the system uploads the model and the movie to the model-sharing Web site automatically.

The system allows multiple downloads, and it will generate slide shows of these downloads. Users can then modify the downloaded models by changing from the play mode to the edit mode with just one click.

7.3.3.3 Model-Sharing Web Site

Modulobe consists of a model-sharing Web site for users to share created models and communicate with other users. Users can upload and download models and add tags and comments to the uploaded models on the Web site.

Figure 7.10 shows a screenshot of the Web site's top page. From the upload form on the Web site, users can upload a model, a thumbnail image or animation, tags,

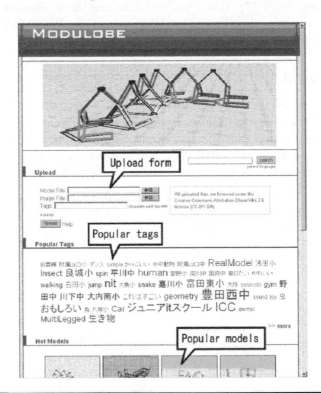

Figure 7.10   The top page of the model-sharing Web site.

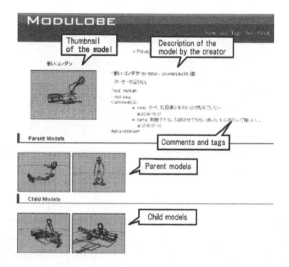

Figure 7.11   Page of a model.

and comments. Users can upload any image file format to use as a thumbnail, but most users upload an animated graphic interchange format (GIF) file, which is generated by Modulobe itself. "Popular Tags" shows tags that many users have added to the uploaded models. They can see a list of models, which have the same tag when they click a tag. In addition, the top page shows a list of the most downloaded models and recently uploaded, tagged, or commented models.

Figure 7.11 shows a screenshot of the page that accompanies each model. This page shows a title and a thumbnail of the model and comments by the user. In addition, it shows tags and comments that have been added by other users. Clicking a user's name shows a list of the models created by this user. The downloaded number is shown at the side of the title. At a bottom of the page, parent models and child models are shown. Parent models are models that are adapted or copied to create this model. The modules of child models are copied partially from parent models. Modulobe can trace relationships among models and display them automatically. Therefore, users do not have to put extra effort in presenting these references.

There are several models that we observed as more popular than others, such as those that move horizontally (e.g., a snake.) Users appeared to indicate that they like this type of models because they could race their models with each other. In a workshop that we held after the release of Modulobe, however, we examined user activities and found that at first users tended to produce a model that moves up and down irregularly. Then they gradually learned to develop a model that can move smoothly horizontally by controlling the motion of links or structures. The creators were learning how to make their models through using the system.

Various patterns to realize horizontal motion have appeared. Some typical models are shown as follows.

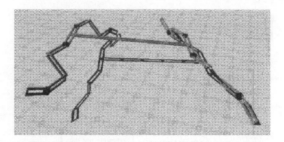

Figure 7.12   A model that walks on its feet.

In the model of Figure 7.12, the right front leg is connected to the left hind leg and the left front leg is connected to the right hind leg. This model moves almost like a buffalo connected to a plow. One variation of this model was uploaded, in which rather than moving like this model, it moved by jumping like a frog.

Figure 7.13 is a model that moves forward by crawling like a snake. The direction of the frictional force between the model and the ground depends on the angle between these two. Therefore, the model can move merely by changing the angles.

Figure 7.14 shows a model that moves by rolling a wheel. This model is very unstable when moving. However, other creators modified this model and uploaded new models with three wheels that can move more stably.

We observed 531 parent-child relationships in the sharing Web site. A parent-child relation is detected when a user reuses a model to create a new model; 349 models (10.4 %) are parent models and 449 models (13.4%) are child models. The parent-child relation can be several generations, but almost all relations are two generations; the maximum is four generations. Figure 7.15 shows the largest parent-child network. A node implies a model, and an edge shows a parent-child relation.

In Figure 7.16, four models are shown (model numbers are 434, 436, 438, and 1148) that have no parent model. Model 436 "changing circle" and 1148 "eight

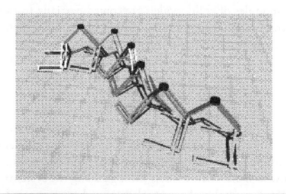

Figure 7.13   A model that crawls on the ground.

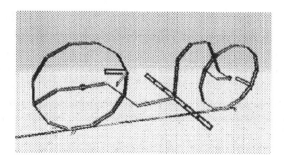

Figure 7.14   A model with wheels.

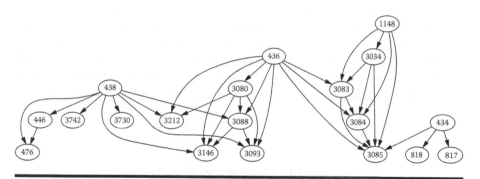

Figure 7.15   An example of a parent-child network.

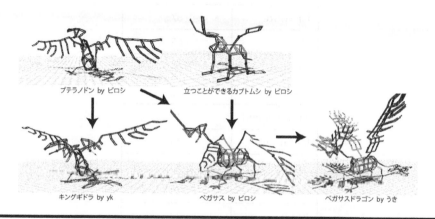

Figure 7.16   Examples of models that have parents and children.

changes" are simple models that consist of approximately ten modules. Most models that reused these models reused other models together, meaning that users use each as components for their own creations, as useful sets of modules. On the other hand, model 434 "super merry-go-round" and model 438 "pteranodon" are complex models, perhaps with more than one hundred modules. Many models reused these models, but most have only one parent, meaning that they use model 434 or model 438 as a base for a new model. One of the heavy users comes to create models as components like wheels, a human body, and a humanoid.

7.3.3.3.1 Adding and Searching Tags

This model-sharing section of Modulobe also allows users to search models using tagging keywords and add tags to a model. Table 7.1 presents a list of the top ten tags (other than user's affiliations.) There are about four hundred unique tags in the site.

Sen classified tags into three categories: "factual," "subjective," and "personal" (Sen 2006). We classified the top one hundred popular tags that are added to models into four categories: Sen's three categories and "system." "System" means a tag that is related to the system and does not have a meaning for users. The model-sharing site does not provide users with a bookmarking function. There are no personal tags. In this case, we classified tags that mean the author's name and author's affiliation into "personal."

Table 7.1 Popular Tags Except for a User's Affiliation

Tag	Models
RealModel	70
Creature	67
Car	65
Human	64
Insect	42
MultiLegged	38
Geometry	38
Interesting	35
Gym	29
Jump	27

Table 7.2   Type of Tags in Modulobe

Type	Tags	Examples
Factual	52	car, vehicle, creature, bird, jump, MultiLegged, spin
Subjective	11	great, interesting, cute, cool, beautiful
Personal (author)	32	Yoshiki Elementary School, tanimoto
System	5	notags, uploadtest

Table 7.2 displays a list of the top one hundred popular tags. The "factual" tags present not only category but also functions and forms of models, e.g., "jump," "spin," and "MultiLegged." These tags are useful for seeking reference models when a user creates a new model. It is difficult to add such tags automatically because the system must analyze the structure and motion of models.

Table 7.3 show the ranking of tags used for the searching model. Sen (2006) described a factual tag as more useful for findings and learning than a subjective tag. However, in this case, a subjective tag is important for searching for a model. The target content of Sen (2006) is a movie. A factual tag is appropriate for helping search if we seek a movie because we know what kind of attributes a movie has. If we cannot predict what kind of content is in Modulobe, a subjective tag is more effective.

Table 7.3   Popular Tags Used for Searching a Model

Tag	Number of search
Interesting	48,059
Great	47,958
Masterpiece	31,492
RealModel	30,362
Cute	27,234
Cool	23,401
Beautiful	18,162
Kagawa EL	16,078
Toda East EL	15,400
Tonda West JHI	14,081

Table 7.4 Popular Models' Tags (Top Three) in the Community

Community	Number of Models	Popular Models' Tags (Top Three)
Modulobe WS	272	insect, RealModel, cute
ICC	172	interesting
nit	127	dance, interesting, jumper
Toda East EL	108	animal, spring, gym
Toyoda West JHI	82	cool, car, creature
Yoshiki EL	74	great, RealModel, cool
Ouchi South EL	74	geometry, interesting
Kawashita JHI	68	animal, string, text
Kagawa EL	59	creature, cute, MultiLegged
Hirakawa JHI	49	car, creature, bird

Note: "EL" means elementary school; "JHI" means junior high school; "nit" is the name of a university.

We held Modulobe workshops with some organizations and events at junior high schools and elementary schools. Each workshop was attended by dozens of participants. In the workshops, facilitators taught participants how to use Modulobe. They learned how to create models not only through instruction from a facilitator but also through discussion with others. Thereby, a creative community was formed at the workshop. Table 7.4 shows tags of popular models of each community. It shows that each community has a different tendency.

7.3.3.3.2 Interaction within Modulobe Community

Because Modulobe provides an easy way to modify other people's models, we observed two types of interactions within the Modulobe community: collaboration and competition. In the former case, users modify an existing model to make it better (e.g., in one case, somebody uploaded a model of bicycle that was unstable. Other users immediately modified this bicycle to fix the problem—see Figure 7.17.) In the latter case, users "got together" and came up with ideas for racing their models.

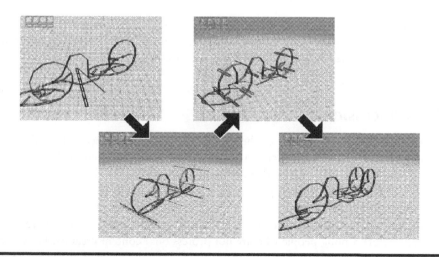

Figure 7.17   Collaborative efforts in improving the upper left bicycle.

7.3.3.4 IP Issue in Modulobe

With the introduction of Niconi Commons, NND had moved in the right direction in regard to crediting the original owner of the creations. However, Niconi Commons still relies on users' goodwill to credit the original content creators. Without enforcement, this voluntary system may not work. Within Modulobe, we developed a mechanism that makes it easier for people to credit (and makes it more difficult for people to infringe on) IP content. The system contains an automatic visualization of the relationship between original content and the evolution of content if modifications have been performed. More specifically, the system supports an inheritance relationship among variations of the same basic content (i.e., when copying or reusing a certain part of the creation is performed.) We also implement a social tagging function in which users can add tags or comments to a model and view the uploaded models from a certain category. We hope that by visualizing the relationship between original content and the evolution of content, users are encouraged to contribute new models more actively. Previous work indicates that displaying the value of contributions stimulates users to work harder (Rashid et al. 2006).

Modulobe handles the IP issue by implementing a function to add a unique ID to all the modules in each model. Every modification to the modules or model retains the original creator's ID and adds the modifier's ID. For example, a new module in model A has model A's ID. The module continues to have model A's unique ID when a user copies and pastes the module from model A to model B. We can find a relationship among models by checking the module ID.

On the Web site, all uploaded models are published with a CC license. Everyone can copy, distribute, and transmit the work on the condition that they attribute the work in the manner specified by the author or licensor. Modulobe has a function to maintain and present such a relationship, and this automatic citation function allows users to more easily comply with the CC license.

7.4 Conclusion

Web 2.0 encapsulates the idea of the proliferation of interconnectivity and inter-activity of Web-delivered content. Arguably, Web 2.0 is the main driving factor behind the rise of UGC. UGC denotes content that is created by general users—which can be in the forms of posts on discussion groups, personal blogs, wikis, and other creative media contributions—with varying degrees of content oversight. UGC sites have facilitated massively collaborative creation activities that we have not seen before among people who are not professional content creators. However, because content is uploaded, modified, downloaded, and deleted at the speed of light, UGC brings with it important issues of copyright and IP rights, partially because some general users are not familiar with these issues (and therefore unaware that they had violated copyright or intellectual property rights [IPR]), and partially because there is a lack of infrastructure to credit the original content owner.

In this chapter, we have described various types of online content sharing and their IP implications. We also presented two case studies: NND, in which users can volun-tarily credit the original owners, and, in more detail, Modulobe, in which the original content creator's ID is automatically inherited when content is modified. Following the development of NND in terms of user behavior, sharing dynamics, and copyright issues, we found that UGC sites cannot fully rely on users' goodwill to credit the origi-nal owners. In July 2008, an agreement was formed for deletion of videos that infringe copyright. Even the introduction of Niconi Commons, which gathers all work under a CC license (meaning all content there is free to use under the terms of CC), the issue of lack of crediting the original owners remains. We argue that there is a need to provide technological infrastructure to automate the recognition of the original author's work, which we implemented in Modulobe. Although the enforcement is in place, users are still actively contributing to Modulobe, which may indicate that for many users, the main issue was not that they were unwilling to credit the original owner; rather, it was an issue of the effort associated with tracking down the original creators and finding a way to embed the owner's ID into the modified creation.

Copyright and IPR play a significant role in content creation, as they encour-age the production of socially beneficial, culturally significant expressive content (Hunter and Lastowska 2004). However, all of the rules that apply to copyright and IP arrangement (and infringement) are undergoing revolutionary decentralization and disintermediation because of the rise of UGC, social computing, and Web 2.0. We close this chapter by suggesting that there is a need to carefully consider the issues related to copyright and IP and to recognize the opportunity and desirability of decentralized content that UGC brings.

References

Ahern, S., Eckles, D., Good, N., King, S., Naaman, M., and Nair, R. 2007. Over-Exposed? Privacy patterns and considerations in online and mobile photo sharing. In CHI '07: *Proceedings of the Special Interest Group on Computer-Human Interaction (SIGCHI) Conference on Human Factors in Computing Systems*, San Jose, California, April 28-May 03, 2007, 357–366.

Baraff, D. 1989. Analytical methods for dynamic simulation of non-penetrating rigid bodies. *SIGGRAPH Comput. Graph.* 23, (3) (Jul. 1989): 223–232.

Boyd, D., and Ellison, N. 2007. Social network sites: Definition, history, and scholarship. *Journal of Computer Mediated Communication* 13, 210–230.

Burton, E. 2008. *Sodaplay.* http:\\www.sodaplay.com\ (Accessed March 2009).

Bush, V. 1945. As we may think. *The Atlantic Monthly*, 17:101–108.

Chadwick, A. 2006. *Internet Politics: States, Citizens, and New Communication Technologies.* Oxford, UK: Oxford University Press.

Chalfen, R. Snaphot Versions of Life. Bowling Green, Ohio; Bowling Green State University Popular Press, 1987.

Cosley, D., Frankowski, D., Terveen, L., and Riedl, J. 2006. Using intelligent task routing and contribution review to the help communities build artifacts of lasting value. *In Proceedings of the SIGCHI Conference on Human Factors in Computing Systems* (Montreal, Quebec, Canada. April 22-27, 2006). R. Grinter, T. Rodden, P. Aoki, E. Cutrell, R. Jeffries, and G. Olson, Eds. CHI'06. New York, NY: ACM, 1037–1046.

Cunningham, W., and Leuf, B. 2001. *The Wiki Way. Quick Collaboration on the Web.* Toronto, Ontario, Canada: Addison-Wesley, Pearson Canada Inc,.

Dourish, P. 2006. Re-space-ing place. "place" and "space" ten years on. In *Proceedings of the 2006 20th Anniversary Conference on Computer Supported Cooperative Work* (Banff, Alberta, Canada, November 04-08, 2006). CSCW '06. New York, NY: ACM, 299–308.

Falco, M. *Springs World 3D.* Falco, http://www.sw3d.net/ (Accessed August 2009).

Hamasaki, M., Takeda, H., and Nishimura, T. 2008. Network analysis of massively collarborative creation of multimedia contents: case study of hatsune miku videos on nico nico douga. In Proceeding of the 1st international Conference on Designing interactive User Experiences For TV and Video (Silicon Valley, California, USA. October 22-24, 2008). UXTV '08, Vol. 291. New York, NY: ACM, 165–168.

Harrison, T. M., and Stephen, T. D. (Eds.). 1996. *Computer Networking and Scholarly Communication in the Twenty-First-Century University.* Albany, NY: State University of New York Press.

Hartman, K., and Koohang, A. 2005. Discussion board: A learning object. *Interdisciplinary Journal of Knowledge and Learning Objects* 1, 67–77.

Holmes, S., and Ganley, P. 2007. User-generated content and the law. *Journal of Intellectual Property Law and Practice* 25, 338–344.

Hunter, D., and Lastowka, G. 2004. Amateur-to-amateur. *William and Mary Law Review* 46, 1026–1027.

Johansson, G. 1973. Visual perception of biological motion and a model for its analysis. *Perception and Psychophysics* 142, 201–211.

Miller, A. D. and Edwards, W. K. 2007. Give and take: a study of consumer photo-sharing culture and practice. In *Proceedings of the SIGCHI Conference on Human Factors in Computing Systems*. San Jose, California. April 28-May 03 2007, New York: ACM, pp. 247–356.

Nakamura, S., Shimizu, M., and Tanaka, K. 2008. Can social annotation support users in evaluating the trustworthiness of video clips?. In Proceeding of the 2nd ACM Workshop on information Credibility on the Web (Napa Valley, California, USA. October 30–31, 2008). Wicow '08. New York, NY: ACM, 59–62.

Nardi, B. A., Schiano, D. J., and Gumbrecht, M. 2004, Blogging as social activity, or, would you let 900 million people read your diary? In *Proceedings of the 2004 ACM Conference on Computer Supported Cooperative Work* (Chicago, Illinois, November 06-10, 2004). CSCW '04. New York, NY: ACM, 222–231.

Nelson, T. H. 1994. *Literary Machines*. Sausalito, CA: Mindful Press.

Nelson, T. H., Literary Machines, Sausalito, California: Mindful Press, 1994.

O'Reilly, T. 2007. What is Web 2.0: Design patterns and business models for the next generation of software. *Communications and Strategies* 1: 17.

Organisation for Economic Co-operation and Development [OECD]. Participative Web: User-created content. http://www.oecd.org/dataoecd/57/14/38393115.pdf (Accessed November 2008).

Osborne, D. 2008. User generated content UGC: Trademark and copyright infringement issues. *Journal of Intellectual Property Law and Practice* 39, 555–562.

Rashid, A. M., Ling, K., Tassone, R. D., Resnick, P., Kraut, R., and Riedl, J. 2006. Motivating participation by displaying the value of contribution. In *Proceedings of the SIGCHI Conference on Human Factors in Computing Systems* (Montréal, Québec, Canada. April 22-27, 2006). R. Grinter, T. Rodden, P. Aoki, E. Cutrell, R. Jeffries, and G. Olson, Eds. CHI '06. New York, NY: ACM, 955–958.

Samuelson, P. 1996. Intellectual property rights and the global information economy. *Commun. ACM* 39, Jan. 1, 23–28.

Schiano, D. J., Nardi, B. A., Gumbrecht, M., and Swartz, L. 2004. Blogging by the rest of us. In *CHI '04 Extended Abstracts on Human Factors in Computing Systems* (Vienna, Austria. April 24-29, 2004). CHI '04 New York, NY: ACM, 1143–1146.

SCEI. 1999. *Panekit*. http://www.scei.co.jp/.

Sen, S., Lam, S. K., Rashid, A., Cosley, D., Frankowski, D., Osterhouse, J., Harper, F. M., and Riedl, J. 2006. tagging, communities, vocabulary, evolution. In *Proceedings of the 2006 20th Anniversary Conference on Computer Supported Cooperative Work* (Banff, Alberta, Canada, November 04-08, 2006). CSCW '06. New York, NY: ACM, 181–190.

Silva, P. A. and Dix, A. 2007. Usability: not as we know it!. In Proceedings of the 21st *British HCI Group Annual Conference on HCI 2008; People and Computers XXI: Hci, But Not As We Know* It-Volume 2 (University of Lancaste United Kingdom, September 03-07, 2007). British Computer Society Conference on Human-Computer Interaction. British Computer Society, Swinton, UK, 103–106.

Surowiecki, J. 2004. *The Wisdom of Crowds: Why the Many Are Smarter Than the Few and How Collective Wisdom Shapes Business, Economies, Societies and Nations*. New York, NY: Doubleday.

Völkel, M., Krötzsch, M., Vrandecic, D., Haller, H., and Studer, R. 2006. Semantic Wikipedia. In *Proceedings of the 15th International Conference on World Wide*, WWW '06. New York, NY: ACM, 585–594.

Waddoups. M. *Juice*. http://www.natew.com/juice/ (Accessed March 2009).

Wan, Y., Kumar, V., and Bukhari, A. 2008. Will the overseas expansion of Facebook succeed? *IEEE Internet Computing* 123, 69–73.

Wiencek, H. 1987. *The World of LEGO Toys*. Harry N. Abrams, New York, NY, USA.

Chapter 8

From Online Familiarity to Offline Trust

How a Virtual Community Creates Familiarity and Trust between Strangers

Paula Bialski and Dominik Batorski

Contents

Hospitality networks are Internet-based social networks of hundreds of thousands of individuals who use an online space to search out accommodation in the home of another network member. Instead of staying at a hotel or hostel, the

179

members of this virtual community prefer to encounter individuals online who are willing to host them in their home. Additionally, users may be motivated to engage in the hospitality network not to travel but to host members from various parts of the world. Other users simply enjoy meeting other network members for a coffee or a tour of a new city. These are networks in which interaction between members originates online with the purpose of meeting offline. Couchsurfing .com is the largest community, with nearly 1 million members globally.

As we will see, trust is central to this community, and trust, as well as the question of trustworthiness, is made very explicit through the design of the Web site. The initial question that instigated the three years of fieldwork with this community was How does trust develop on this sort of Web site? What enables strangers to trust one another, and how is it possible that a community enables trust between users both online and offline? After speaking to over thirty community members, conducting an online survey (which reaped around three thousand respondents), sleeping on couches, floors, and mattresses from Montreal to Stockholm, and hosting over thirty Couchsurfers, we came to the realization that trust is multicausal and is dependent on a user's perception of familiarity. It was Luhmann who believed that "Trust has to be achieved within a familiar world ... we cannot neglect the conditions of familiarity and its limits when we set out to explore the conditions of trust" (Luhmann 1979). The central question in this chapter is how does a virtual community such as Couchsurfing.com create this "familiar world," which then creates trust between users? We will tackle this question by explaining that trust emerges in the following stages:

(a) First, individual predispositions enable a person to relate to the purpose of a given online community. Only those who relate to Couchsurfing.com and its purpose—hosting or visiting strangers in one's home—join the Web site and use it actively. Thus, a process of self-selection occurs, and Couchsurfing users usually have a more open approach to trust than those users who would not join such a Web site.

(b) Trust is also established through navigation through the online profile, which allows a user to become familiar with another user. The most important elements of a profile are a user's self-presentation and the recommendation system where other users rate their relation and experiences with the user as a host/guest. This user's self-presentation enables others to create a sense of similarity with them, which facilitates the emergence of trust through the mechanism of homophily. The opinions of others in the recommendation system also influences the trustworthiness of a user.

(c) Finally, trust is strengthened during offline contact, during interaction within a user's home while couchsurfing.

To help us explain these processes, Dieberger (2003) introduced a useful concept called social navigation. A much broader concept than recommendation systems, the "goal of social navigation is to utilize information about other people's behavior for our own navigational decisions (p. 35)." As we will see, Couchsurfing serves as an example of a space where users must create a sense of familiarity with other users in order to trust them enough to agree to hosting or visiting them. This chapter argues that users gain a sense of familiarity through social navigation, moving through information about other users based on their personal profiles. But in order to understand how a virtual community creates this "familiar world," we must take into account both how a user socially navigates through a Web site like Couchsurfing and how the Web site is designed in order to enable users to better communicate familiarity between one another. Thus, in addition to the process of social navigation, in this chapter we will discuss the design, or affordances—properties of the environment that offer actions to those wanting to function within the environment (Gibson 1979), offered by Couchsurfing.

In specifically focusing on the way a virtual community fosters trust among its members, this chapter will provide an ethnographic account of this online–offline hospitality network. Couchsurfing is a community of trustful relationships between strangers—ones where interaction moves from the virtual to the corporeal world. Ethnographic description will be employed in order to understand the microprocesses that help create familiarity between users using Garfinkel's concept of "perceivedly normal environments," which he believes is related to trust. "To say that one person 'trusts' another means that the person seeks to act in such a fashion as to produce through his action or to respect as conditions of play actual events that accord with normative orders of events depicted in the basic rules of play." (Garfinkel 1963, p. 193). Thus, to understand the way trust is formed, we will take into account.

(a) the affordances for interaction made by the Couchsurfing Web site,
(b) the way the user socially navigates through the online world and offline world,
(c) the "rules of play," which cement themselves as normative patters of interaction within this virtual community, creating a "perceivedly normal environment."

This description will move from the moment a new member is introduced to the virtual community to the stage of complete immersion. The practice of Couchsurfing will also be outlined in order to understand how a sense of familiarity is created and how trust is then embodied and exchanged in both an online and then offline setting. The description will be narrated by using one user—a woman from Finland—who will serve as a representation and at times a conglomeration of the experiences expressed to us by Couchsurfers interviewed. By describing her experiences, we present an individual who, in our opinion, expresses the actions and opinions of a typical Couchsurfer. Nana experiences, as well as the other interviews featured here, were conducted throughout the course of our three-year study of the Web site. Most of the interviews took place in Montreal between July and August 2006 among Couchsurfers who had been actively hosting or visiting other members of the community. Some community and Web site statistics, collected in November 2008, which are provided to the public through Couchsurfing.com will also be used. Additionally, we will draw certain information from our own online survey, which was conducted between August 2006 and March 2007. This online survey was programmed into the Web site and was made available to users through their individual profiles by adding an extra tab named "My Survey" onto their profile.

8.1 Inside Couchsurfing

The knowledge of an online community such as Couchsurfing often travels in two ways: first, through the media, and second, through one's personal network, often through weak-ties (Granovetter 1973), such as, acquaintances and people we meet through casual meetings, meetings-in-flux, and in public spaces, such as, airports, hostels, bars, and sports clubs. Couchsurfing.com is a talking piece. It is a community of individuals who use the online social networking Web site in order to find accommodation in the homes of other members of the network. It is quirky. It sounds risky. It seems like something your mother would disapprove of. If you have not heard of Couchsurfing, you will certainly want to hear about it. It has slowly entered the small talk of the rich, northern societies—an icebreaker, a piece of small talk, a note for the back page of a life-style magazine.

Nana is an information technology (IT) consultant from Helsinki. She likes her job, she earns enough money to live comfortably, and she is still in her late twenties. She likes meeting new people but complains that she cannot really find anyone to "connect with." "The Finnish are quite closed people," she explains. One day after work, she goes to a pub with some of her colleagues. She knows three out of five of the other people. A friend of a friend starts telling them a "funny story" about his last city break to Berlin, where he went "Couchsurfing." Some have vaguely heard of Couchsurfing, also through word of mouth, through friends of friends, or through loose ties. The new acquaintance begins to explain the nature of the community. Instead of staying at a hotel or a hostel, he stayed on somebody's couch. The group gets excited. This idea seems too good to be true. Questions and skepticism start flying. Did you know them? No. So they were complete strangers? Yes. Weren't you scared that they would hurt you? Not really. Why? Why aren't you afraid? Nana does not say anything and listens to the questions. She is not afraid of the idea. The concept intrigues her. A cheap place to stay. The feeling of home in a foreign place. And insider's view on a place. This is so simple; why hasn't anyone thought of this earlier, she asks herself. Nana is like most Couchsurfers we interviewed—they feel differentiated; they define themselves as people who follow a different set of rules; they believe that "Couchsurfing is not all that radical, it's just different." A 52-year-old woman Couchsurfer explained that "Couchsurfing is a community. People on the same wavelength, people who maybe have the same political views, the same views of the world, the same views of friendship, the same views of trust, and the same views of travel." This is symptomatic of all those who join the community—they just "get it." To a certain degree, not everyone can be a Couchsurfer. Despite the fact that membership is open to everyone, users usually have certain individual predispositions: they must be able to accept the Couchsurfing ideology, including an openness to people they do not know. Moreover, they must believe that people are inherently good natured. This process of self-selection to Couchsurfing makes the emergence of trust between members of this community much easier.

Couchsurfers who engage in the community are looked at as outsiders to the rest of society, and they do not mind being outsiders, being "different." The idea of Couchsurfing, much like Nana experienced among her acquaintances in the pub, is often rejected, thought of as *strange*. But Nana wants to "expand," wants to be different, wants to be "strange" because she is not satisfied with the status quo. Another Couchsurfer explained the people around him in Holland as "being so static, as they are, they didn't get out. Even though we spent all these years together as friends, drinking together, or whatever, they still stay static; they're still in the same place, mentally too." This sense of being static, of immobility, is exactly what many Couchsurfers wish to break free from. Michael, a twenty-eight-year-old American stated, "People are quick to take things that they don't understand or are very different to them and just kind of push them off. And I think Couchsurfing gets that, wanting to expand relationships in a more free way." For Michael, this is a freeing, liberating practice. The understanding among Couchsurfers is that while "mainstream" society associates interaction with strangers as something involving risk. Couchsurfing as a practice adheres to its own set of rules of friendship and familiarization—and Couchsurfers perceive this to be an alternative to the mainstream. The users express relief at the "ease" in communication with other Web site members and the "common understanding" that generates a sense of closeness. Nana stated that communication is "easier with Couchsurfers because there is some common element, some kind of spirit that combines you, that's not seen around the people I'm usually around in my place … Couchsurfing is a heightened reality."

That being said, Couchsurfers are often drawn to the idea out of either an understanding of the practice (i.e., they have created informal networks of hospitality where they used to stay on the couches of friends of friends or hosted people they met in a pub or at a party) or a desire to engage in a practice that promises an alternative to the risk society (Beck 1992) under an individualistic, consumer capitalism—it is an utopian altruism. In other words, those like Nana who first hear and wish to engage in the practice desire to *become trustworthy*, and Couchsurfing is their ticket to engage in such a practice. This desire is created as a dichotomy to the social reality of a risk society. As Nana stated, it becomes the heightened reality—one where trust between individuals is the norm, not the exception. Self-disclosure and openness create a level of psychological risk, yet members of this community create an illusion of close disclosure by building up trust between each other. Trust allows users to create closer relationships despite the risks involved. Human relations are not purely symbolic as Baudrillard explained (1994), a simulation of something real; here they are real—they are face to face, tangible, intense, and they promise intensity. Moreover, although we could come to the conclusion that Nana wanted to become part of Couchsurfing because she has the time, financial resources, and accommodation to host another individual, this is not often the case. Couchsurfers have many faces. They can be families, single men living in bachelor apartments, young mothers, students living in residence, or farmers. Calling "Couchsurfing" by its name is deceptive—visitors sleep in tents, on the floor, on blow-up mattresses, on children's beds (the children then sleep with their parents), in a barn, or sometimes share a bed with the host. Thus, the individual's lifestyle is irrelevant. If a person individual is interested in the idea of Couchsurfing, she will arrange her life around the practice. The main slogan that appears on the Web site is a call to "Participate in creating a better world—one couch at a time." So in order for Nana to become engaged in the Web site, she must identify herself with the ideology behind it. The bare-bones structure of this type of exchange promises new users a cheap way to travel—a free couch. Yet Couchsurfing aims to weed out this type of utilitarianism and provide instead an ideology with a free couch: "This is not just a free couch. This promises intense, frequent, diverse interactions," stated the founder of the Web site. When Nana has identified with the ideology, she will expect that the practice that the Web site promotes is interesting, good, safe, something she also wants to engage in. Another Couchsurfer stated that by being a member, he "looks forward to lots of enriching experiences, stimulating conversations and fascinating people." This expectation of a worthwhile venture is, in essence, a trust that this practice is functional. Specifically, one trusts that through Couchsurfing, one will indeed find "fascinating conversations" and "stimulating people." It is the ideological promise of a better reality that this given online community expresses that drives one's desire to become a member.

8.2 Online Community Ideology

When a community shares common ideals or goals, a basic level of trust is always present. Lewis and Weigert identified this type of trust as a "social reality." As such, individual's interactions with other community members cannot be simply construed as being trustworthy or not. Specifically, within a community, individuals are not faced with the hyperbolic extremes of "trust" and the "absence of trust," decisions that are cognitive and present outside of such a community. In groups for which trust exists as a social reality, interpersonal trust comes naturally and is not reducible to individual psychology (Lewis and Weigert 1985). Couchsurfing is based on a common purpose and ideology, where members believe that opening up one's home to strangers will provide various cultural, educational, and self-reflective benefits. What is also worth noting is that trustfulness becomes a mandatory practice in order to become part of this community. In other words, you cannot be a Couchsurfer if you do not want to trust another person enough to let him into your own home, your own private space.

Nana adheres to this ideology. She does not care if a "stranger" sits on her sofa. Her sofa is disposable, the experience is lasting. She is attracted to the experiences that the membership to this privileged community will potentially provide her—the chance to interact with new people, the chance to be open with others, the chance to share. The new sights, the new smells of entering a stranger's private space attract her, as well as the chance to travel to places too expensive for her. A single mother in her early thirties living in Leipzig mentioned that Couchsurfing is attractive because it allows people into the "private sphere [which] has all sorts of emotions attached to it. And when I share that space with someone, it just feels more natural to be emotional and honest, and have an honest discussion. Outside, we have small talk. And small talk is not me, it's not about me. It's about nothing. But these discussions are anonymous because the surroundings, too, are anonymous." These are all new chances, new experiences, something she desired to live through.

Among her group of friend in the pub, Nana cannot articulate why she does not think Couchsurfing is a bad idea. She just writes down the URL (uniform resource locator, i.e., the Web site's address), and they move to another topic. Because Nana identifies with the community's ideology, she does not experience a sense of risk or discomfort when introduced to the idea. This is the first mechanism of trust that allows this community to function; trust is a social reality, part of the ideology and practice of the community.

The desire to be part of this utopian ideal, or the urgency that this sort of utopianism exists and one is not part of it, drives the new member to join the community soon after they first hear about it. Nana is now logging onto Couchsurfing. Although earlier I mentioned that the living patterns of Couchsurfers vary, those who gain the desire to join the Web site must first have an understanding of what a virtual community is and how it functions, and, quite basically, they must be able to log onto a Web site. Those who desire to join Couchsurfing must understand and trust the system that is the Internet. Without already holding a narrative of the Internet, the belief in the functionality of this type of virtual community is impossible.

8.3 Creating an Online Profile

The process of self-presentation online is an important aspect in relational development offline. Couchsurfers must establish an online profile of themselves, and this profile communicates information about oneself that is crucial in creating a sense of knowledge about the other person, which influences familiarity. These features in the profile are what we will call "affordances"—properties of the environment that offer actions to appropriate organisms (Gaver 1992). The Couchsurfing profile offers certain affordances that will be discussed, which influence the way a user is perceived by other community members. The profile also plays a crucial role in a user's social navigation through the Web site, when she chooses whom to initiate contact with. Goffman's work on self-presentation explains that certain individuals may engage in strategic activities "to convey an impression to others which it is in [their] interests to convey" (Goffman 1959). As the average age of Couchsurfers is twenty-seven, and 70% of users are from Europe or North America, most new Couchsurfers are like Nana: They have a high level of media literacy and are proficient at navigating through Web sites. It takes her a matter of minutes to register her profile. Communicating trust is linked to the way we communicate "self" to another user—directly helping the other to familiarize himself with who we are. This virtual community provides a number of affordances for its members to communicate a sense of trustworthiness to another community member. Nana fills in a variety of boxes that ask for her "personal interests" and "life approach." She types in that she likes Finnish literature. Her interests are "strange lands and languages." Couchsurfing asks her to fill in her "current mission." She writes: "To have brand new adventures." She is asked to explain the "types of people she enjoys." She writes: "Seekers. People who are curious, creative, laid back, insightful without being pretentious and intelligent without being arrogant about it. Those who prefer shaking the world with their thoughts rather than just making noises."

These profile questions are crucial in yet another mechanism that promotes trust within this virtual community—communicating a sense of familiarity. Dan, one of the founders of the Web site, explained that the profile questions are structured in such a way that "it brings out the essence of people. And when people's essences are visible, it contributes to the building of trust." This makes us assume that (1) trust is dependent on the level of self-disclosure, and (2) trust is dependent on the levels of homophily (similarity) between users. Therefore, if a virtual community like Couchsurfing wishes to create trusting virtual relationships, it must provide enough space for the user to disclose who they are to the rest of the community. When someone like Nana builds her first profile, a Web page flashes in front of her, suggesting that the more information a user provides, the more trustworthy she will appear to other users. The Web site also suggests providing some personal description. In doing so, "those viewing your profile get a better idea of who you are. This helps others understand who you are based on the content of what you write, and also how you write it."

According to Web site statistics, 60.1% of the members present photos, and most active members have a number of photos attached to their profile. Profiles with text and no photos can be considered less trustworthy, and users are encouraged to post a photo online in order to get better feedback. One of the main aspects, which make Couchsurfing distinct from all other Web sites of its sort, is its high use of photo images. Unlike other virtual communities like Facebook, the photos attached to the profile must be actual likenesses of the members themselves. And although this does not guarantee that a member will not use somebody else's headshot, photos of celebrities, animals, or objects are never used and are deleted by the Web site administrators.

Couchsurfing also features a "buddy list" or a "friendship list." Every profile shows a list of other members that the given person knows. As an example of how the friend list works, let us consider the profile of Andrew in Warsaw, Poland. He has over 180 friends linked to his profile, and most of them are people he has strictly surfed or hosted, although they also include high school friends, university friends, and family members. Every time somebody clicks on Andrew's profile, they can look at all of Andrew's friends, and all their "references" for Andrew as well as their "link strength" (which we will explain in detail below.) By using simple common sense, a user will come to the conclusion that somebody with many friends is, in fact, more likely to be trustworthy. It is also worth mentioning here that the type of person one is friends with matters just as much as the number of friends one has. For example, a person who has all of the Web site administrators and a few ambassadors as their friends could also be considered more trustworthy.

Members who become very involved in the development of the Couchsurfing project, want to help promotion of Couchsurfing, or are highly active surfers and hosts have the potential of becoming a Couchsurfing "ambassador." The ambassador's job comes in all shapes and sizes. Sometimes an ambassador can help translate the Web site into a new language or send "greetings" to new members. The title of "ambassador" is given to a surfer by members of the administrative team, which is a collective of individuals who helped found and develop the Web site. The fact that a person is an ambassador, or a founding member, has the potential to make him more trustworthy.

Each member has the opportunity to enter the process of vouching. One member can "vouch" for another member by clicking on an icon next to her profile. In doing this, one member makes a sort of promise to the rest of the Couchsurfing community that the member they are vouching for is, in fact, trustworthy. Often people can vouch for their good friends or for people they have hosted or surfed with. New users like Nana obviously cannot vouch for anyone if they do not have anybody linked to their profile. Vouching implies that you have been a member for some time.

Verification is something slightly different from vouching and involves a four-step process in which a member sends his credit card number to the members of the administrative team, the founder of the Web site. The members of the administrative team in charge of verification then take $25 U.S. from the member's account (there are discounts for those who cannot afford the fee.) Upon payment, the user is sent a verification number to her home address. The member enters the number onto the Web site and receives the full level of verification. Verification guarantees others that this member lives at the address indicated on the Web site and that her identity is correct. The downfall of this process is that it excludes those without a credit card and financial means to pay for it. Verification is not a requirement of the Web site, and many members visit or host other users without it.

Relationships on Couchsurfing are recorded with a set of variables and descriptions, meaning, when Nana adds a friend to her "friendship list," she must answer a set of questions about this friend. These questions are based on the origin and duration of the relationship. From a drop-down menu, the user chooses from a list of thirteen answers about the relationship origin. Nana would also have to explain how strongly she trusts the friend and would have a choice of six text-based answers that range from "I don't know this person well enough to decide" to "I would trust this person with my life" (other answers include "I don't trust this person," "I trust this person somewhat," "I generally trust this person," and "I highly trust this person.") This friendship list is an indicator of how popular an individual is and, through the "trust strength," she gains information on how reputable (and hence, trustworthy), she is.

The final feature, and one of the most important on Couchsurfing explicitly relating to trust, is the reference system, which is similar to online auction Web sites like Ebay.com, where a user is given a positive/negative rating from people who have interacted with him either through hosting or visiting, or elsewhere. This feature functions in addition to the friendship list, and under any Couchsurfer's profile there can appear a list of references from old school friends, family members, or former hosts and guests. Note that a reference is usually left by someone whom we have met and interacted with face to face, over a period of time. One of Nana's references written by her former host was

> Another Couchsurfing enthu(siastic) person. [Nana] is one of the first people I hosted through Couch surfing ... hosted her in Bombay with her friends from England, then met up with her in Goa and traveled a bit together and again hosted her at my place ... It would be funny if [Nana] wrote a mail asking me whether she can surf my couch ... she has become family now ... the wanderer spirit in her and in me will ensure that we meet some day in some corner of this not so lonely planet and I am really looking forward to it.

None of the received references can be deleted by the user, and some profiles include negative references. According to Web site statistics, 99.8% of experiences are positive, but the negative experiences are also posted on a user's profile. These references are textual and depend on the detail; a lot of information is contained in the descriptive opinion of each user. The level of enthusiasm can differ from user to user. The meaning in these references cannot be understood by reading just the "positive" or "negative" rating, yet rather by reading the detailed description of the author. Knowledge of this reference feature may also implicitly encourage community members to be good hosts or guests during the offline interaction so that they may gain a positive reference and, hence, a positive reputation online.

In general, the profile serves as an introduction, and Couchsurfing acts as the mediator in this introduction, helping acquaint one person with another, offering various affordances within the profile by which a member can communicate who they are. Yet this is not a regular "acquaintance" in the traditional sense of the word. Georg Simmel, who wrote extensively on the subject of socialization and interaction, believed that mutual "acquaintance by no means is knowledge of one another; it involves no actual insight into the individual nature of the personality. It only means that one has taken notice of the other's existence, as it were" (Simmel 1949, p. 320). Yet, Simmel, writing at a time when online socialization was nonexistent, did not take into account the virtual phenomenon of online profiles—when strangers could gain information about one another through their online identities and, before meeting each other in person, create an illusion of closeness and familiarity. In order to relate this to the process of social navigation, let us consider that just as "The number of cars parked in front of a restaurant is an indication of its popularity as is the length of the waiting line before a theatre" (Dieberger 1997, p. 807), the number of photos, friends, references, travel experiences, etc. can have a deep impact on the way in which we perceive another user, as we will see in the following paragraphs.

8.4 Choosing Whom to Trust

After Nana builds her profile, she searches couches to stay on. Nana wants to visit Sweden for the weekend. She types "Stockholm" into the search engine, and thirteen pages showing twenty-five users per page appear. These profiles are smaller versions of the larger profiles and include information such as the user's photo, her age, gender, location, date of last login, languages spoken, hobbies, and "philosophy," which is a short sentence describing the user's approach to life. The number of photos and friends and her email response rate are also shown.

Despite Nana's sense of trust in the ideology of Couchsurfing, some first-time users want to understand how to communicate trust online. A section of the Web site is devoted to answering questions regarding first-time interaction between potential host and guest. The following is an extract from a page found on the Web site, outlining the precautions new Couchsurfers should take when hosting or surfing.

> When you receive a Couch Request, try to get an idea of who is writing to you. Is it clear that they have read your profile? Are they interested in meeting YOU, or does it seem like they are just looking for a free place to stay? Who are your Surfers? A profile can give you valuable information about a person, so first find out what kind of people will surf your couch! Read the profile carefully. Is their profile filled out? Do they have photographs? Who left them references? How much CouchSurfing experience do they have? Did that glowing reference come from someone who hosted or surfed with this person? (Couchsurfing.com, October 17, 2008).

The fields that are filled in by each user not only allow Nana to familiarize herself with another user but also allow her to discern whether or not the other user would be a person she wishes to interact with. This decision is quite subjective and often based on homophily—or an attraction to those who are similar to Nana. Among all the users in Stockholm, Nana found a woman her age who, like her, is interested in creative, adventurous, and open-minded people. Similarity instantly connects Nana to this stranger, and in this case, this sense of connection is based on the stranger's perceived worldview. This user in Stockholm has now shifted from being a stranger to being a familiar, like-minded individual—all before any verbal dialogue was exchanged. The design of Couchsurfing enabled this connection to take place.

Despite the range in demographic profiles, as Nana logs into her account and chooses a couch to stay on, she will be more likely to visit someone who has the greatest similarity to her. Ziegler and Golbeck (2007) explained that dependencies between user similarity and trust exist when the community's trust network revolves around a common goal or particular application and when the individuals trusting each other share similar interests or traits. Trust, therefore, is positively related to homophily. Let us take into consideration an analysis carried out on 221,180 friendship dyads registered on Couchsurfing in February 2006, which includes various variables such as the origin and the duration of acquaintanceship as well as how strongly the individuals trust one another on a scale of 1 to 6. In Figure 8.1, we notice that for all factors, trust is significantly higher within homophilus relations than within heterogeneous ones. For example, the more similar the dyads are in age (no more than a two-year age difference), the more likely they are to trust each other. Specifically, it is evident that if the dyads are of exactly the same age, they are more likely to highly trust each other. 23% of same-age dyads gave each other the highest possible degree of trust, compared to 17.9% of similar age dyads and only 8.8% for dyads with a more than two-year age difference.

A higher tendency toward trust is within homophilious relationships according to age and country of origin, and a slightly lower tendency toward trust is found in same-gender relationships. The same city homophily analysis shows those dyads who know each other from an offline setting. We therefore are not able to decide whether the observed result is an effect of homophily or an effect of the context of a given relationship. These dyads might have had more opportunity to strengthen their relationship offline, in a variety of contexts, compared to other relations who only know each other through the Couchsurfing system. However same-country dyads can, and within our study do, show a tendency for homophily. Our results show that if dyads share the same country of origin, they are more likely to trust each other. This can be linked to a feeling of national identity or the perception of gaining a common understanding that can be linked to a lower sense of risk that facilitates higher trust. Therefore, we can state that people in homophilious relationships tend to trust each other more than those in heterogeneous relationships.

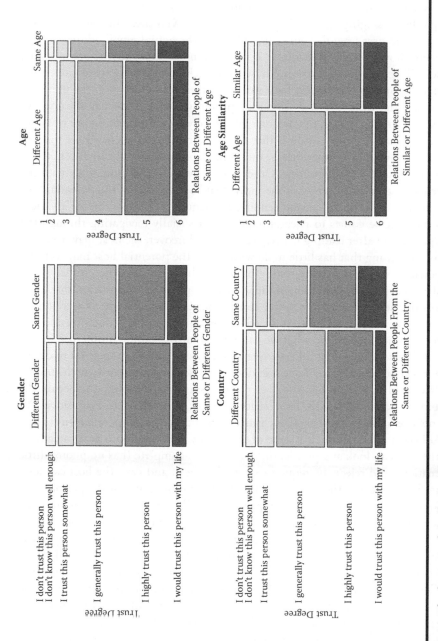

Figure 8.1 The impact of similarity of gender, age, and country of origin on levels of trust (the width of bars is proportional to the number of relations of each category.)

8.5 Making First Contact

Once Nana has found someone she would like to visit, she sends him an email through the messaging system that is built into the Web site. This message is forwarded to the user's regular email provider; yet the actual email address of any user is not shown on the Web site. It is not guaranteed that the first person Nana chooses will accept her as a visitor. Those who receive email requests to Couchsurf may not respond to the request for a variety of reasons: Sometimes, the user may not have any free time to host a visitor that specific weekend. They may already have other Couchsurfing guests, or any other guests, staying with them on the weekend Nana requested. Another reason is that they may not have the energy, or the will to host, as hosting can sometimes be quite time consuming and can also take the host away from their daily routine. Other reasons, which have more to do with the user requesting to visit than with the individual hosting, is that the potential host may not want to host the person sending the request. This may be for a variety of reasons, but he comes to reject a request after socially navigating through the user's profile and after reading her request email. Moreover, the host may reject the user for something that has little to do with trust: The potential host may feel that this potential visitor is not interesting or compatible with her. The profile, which served as a partial representation of the individual, provides information that helps the host or guest come to a cognitive conclusion regarding who he is as a person. Although the amount of honesty in self-presentation online is questionable and has been studied elsewhere, what we can say is that the online profile is the only method the users have to familiarize one another with who they are prior to deciding to meet one another. So, if one user does gain a sense of familiarity with another user after reading her profile, she does not necessarily have to like the person she has come to be familiar with. Familiarity does not always have to be positive. We must not overlook the fact that Couchsurfers do not always accept one another into each other's homes. Thus, this sense of familiarity is divided into two categories: one, the host can look at a profile, think that it is incomplete (has no photos, little information), and reject the request to host the guest; and two, the host can look at a profile, think that there is quite a lot of information, but that the person himself does not seem to be her "type of people." Emails requesting a couch are often sent out to a number of people in a given city. Many respondents and friends have expressed having difficulty finding accommodation in certain cities. Among all the reasons provided above, this also has to do with the time of travel as well as the number of Couchsurfers in a given city. Yet Couchsurfers have expressed the sense that Couchsurfing is about choice: The profile offers them the ability to choose one member over the other, and the host chooses the visitor (and vice versa) based on the perceived ease of interaction. Specifically, Couchsurfers aim to minimize the high risk of social awkwardness when the two go from interacting online to meeting offline. Thus, the profile helps (a) a guest or host to express who she is, and (b) allows the host or guest to discern if the given person will be someone he wants to interact with.

8.6 Trust and Expectation

Yet as Sztompka suggested, nobody knows how someone will act in the future. "If we had such full possibility of prediction, and strong, certain expectations, trust would be irrelevant. Trust is a bet about the future contingent actions of others and involves specific expectations" (Sztompka 1999). Once the host accepts the guest, only a few emails are exchanged—mainly emails providing directions that sometimes include a short informal description of the host's or guest's expectations. Traditions of how to act as a guest in a host's home depend on one's cultural background as well as on one's social awareness. This means, that despite the fact that someone has been taught not to open the fridge and take a host's food without asking, his level of social awareness makes him adhere to these norms. But despite one's level of social awareness, a difference in behavioral norms based on one's cultural background may create misunderstandings between host and visitor. The founder of Couchsurfing, a 30-year-old Alaskan computer programmer, expressed his concern with the community's potential for cross-cultural misunderstanding when visiting or hosting. Based on this concern, the administrators of the Web site created a field on the user profile titled "Couch Information" that "will give surfers information on what they can expect and also what might be expected from them upon their arrival." Dana, a 50-year-old American, is an experienced community member, having hosted more than 50 Couchsurfers on separate occasions. Under her "Couch Information," she outlines a set of expectations.

> I have a guest bedroom with bath available for one or two people, 2 couches, one quite small, and a little travel trailer. I have plenty of bedding. I live alone with my dog Luna about One and a half miles from town. The park is about half a mile so you can walk or ride bikes (I have 4 extras) to town or through the beautiful park. Smoking is allowed outside. I work from home so most the time I will have plenty of time to show you around. If you stay more than two days it would be great if you chip in for food and drinks if you want to have dinners here. I love to cook and you won't regret it! Please let me know at least a week in advance so I can Plan for your stay.

Note that this feature in the user profile provides Dana with space to express her expectations and allows the potential guest to understand what Dana's expectations are of them. If the guest disregards these expectations and, for example, smokes indoors, Dana will lose trust in her guest. A set of expectations provides the guest with an implicit understanding that he will be considered trustworthy if he follows the outlined rules. The guest also avoids any cultural misunderstandings by adhering to the rules set out by the host. As stated earlier, another way expectations can be communicated online is through the messaging system. Nana's host may explicitly write her though the Couchsurfing email system that she (host) must study on the weekend and can only spend time with Nana in the evenings. This provides Nana with a clear set of expectations.

There are two mechanisms of expectation at play during the host-guest interaction. Hosts hold *explicit expectations* for their guests, and these expectations are made explicit through in-profile descriptions such as Dana's. Couchsurfers also have *implicit expectations*, hopes, or desires for the host or guest. The host, for example, may hope that her guest will bring her a gift, or the guest may hope that his host will show him her city, spend time with him, engage him in fascinating discussion, or bring him to an underground party. The implicit practice in Couchsurfing is co-present, face-to-face interaction between the host and guest. When the host's or guest's implicit expectations are unfulfilled, this results in disappointment rather than lack of trust. The unfulfilled implicit expectations are indirectly linked to the development of trust, as the complete development of closer relationships will not take place if the host and guest do not get to know each other through the expected interactions. Yet more directly linked to trust is the breach of an explicit expectation. When a Couchsurfer does not comply with an explicit expectation and, perhaps, smokes in the host's apartment despite the fact that the host's profile requested nonsmoking guests, the guest becomes less trustworthy in the eyes of her host.

8.7 Meeting Nonstrangers

When Nana travels to Stockholm, she is not fearful; rather, she is interested to see whether the familiarity and connection that she anticipated after reading the online profile of her future host are, in fact, present. Couchsurfers express that they do not sense fear when meeting their host or guest for the first time, rather, hold a sense of curiosity. This is yet another way in which trust over Couchsurfing functions: Nana trusts that the other Couchsurfer will be the person Nana expected her to be through the way her "self" was communicated in her online profile. To use Sztompka's phrasing, trust is a sort of bet about the future contingent on actions of others (Sztompka, p. 25). And much like the excitement that comes in winning a bet, Couchsurfers are rather curiously excited to find out if they "won" and were correct in their expectations. Couchsurfers often meet in public places for the first time. This has to do more with cultural norms that have been established than with a perception of risk or distrust in the other person. As an example, if Nana or her host states that she wants to meet in the host's apartment, she might perceive the other as being too straightforward and careless, and she would have reason to be suspicious. On the other hand, if Nana or her host suggests that they should meet in public first, one of them might perceive the other as being too paranoid or suspicious of her and be turned off by the other.

So despite the fact that Couchsurfers believe their interactions to be "more free" than the norm, Couchsurfers are still members of a society and have been socialized into a set of social rules and norms of how and how not to interact with people one has not yet met face to face. These sets of norms standardize the practice of even something so "free and open" as Couchsurfing is perceived to be. Despite the proliferation of online virtual communities, skepticism still exists when interaction moves from an online to an offline space. Suspicion is linked to an individual's perception that another member of society is not playing the social game correctly. Thus, in order to enable smooth interaction, both the host and guest risk little by offering to meet in a public space.

And although Couchsurfers adhere to certain societal norms, there are other rules involved in the practice of hosting or visiting a member of a virtual community in an offline space, which must be established by the individuals involved. Adam, a 40-year-old American, when discussing his first interaction via Couchsurfing explained that "for both of us, it was our first time. So we didn't really know [what to do]." After the host and guest meet and exchange greetings, most often the host leads the guest to his home (although sometimes the host has prior engagements, and she gives the keys to the guest and directs him to her home or simply takes him to her engagement; this is less common.) Adam explained what he told his first Couchsurfer, who ended up staying for ten days:

> "Well, so there you are. You're going to be here for a couple of days … ok well, so I guess you'll sleep here, I have a mattress for you. What are you doing tonight? Tomorrow? Hey you know what, I'm going to a party tomorrow night? And I have another one this weekend, and I have this show I'm going to go to. And there are a few people who want to meet you …"

The tourist ritual of seeing a sight, the collective conventionalized belief that one object or space must be experienced visually is central to the traditional idea of tourism yet completely secondary to this sort of travel that is experienced through the hospitality network. Adam's guest's motivation to travel was not to have a sight-seeing experience in the traditional sense of the word. In our online survey out of eight different possible responses, only 14% of respondents answered, "seeing interesting sights of the world" as their primary motivation to travel, whereas 56% of respondents chose "personal growth/personal development (learning about yourself and the world around you.)"

We can note that the daytime or evening activities are dependent on the host's lifestyle, and frequently the host and guest stay in the vicinity of their host's home, immersed in dialogue. 42% of respondents in the same online survey expressed spending more than three hours per day with their hosts or guests, with 16% spending over six hours per day with them. At the same time, 28% said they spent between three and six hours and 20% said they spent between one and three hours per day. In this sense, hosts attempt to involve the guest in their own lives as much as possible, but the level of involvement is often established online prior to meeting. If Nana had an uninvolved host, she would have to be more independent, occasionally asking her host for directions, maps, or sightseeing tips.

A Couchsurfer like Nana may stay with a host usually between two days and a week. Having only a limited amount of time, the guest and the host will maximize this time. The regular process of "getting to know" one another is quite different and extremely condensed compared to a friendship-making process where two people are grounded in a given locality. The process of familiarization between host and guest within this community functions on a different temporal continuum; both the time and the norms usually found within friendship shift due to various factors, which will be discussed. The first Couchsurfing visit of Karen, a 27-year-old Australian, was to a paramedic in his late thirties who lived in Dublin. Karen told me that the first night she stayed at his house, they had a conversation until two in the morning. When asked what they talked about, she explained: "For him it was the death of his wife, how he felt about his wife. Being diagnosed with polycystic kidneys. Kids ... relationships with his siblings. Travel stories. What his disappointments in life were. What he likes about life. What he liked about Northern Ireland. Some really interesting stories he'd seen because he'd been an ambulance driver during the troubles. Really interesting stories about um...shootings and things like that. He was totally open to any question I'd ask him." And Karen said that she "had the same with just about every Couchsurfer."

This level of self-disclosure and sped-up process of familiarization can be directly linked to the process that was started online. The design of the Web site instigated a sort of interaction between strangers, one where self-disclosure through one's profile was the initial method of contact. When two individuals meet in a public setting, their initial discussion is rarely based on the type of self-disclosure expected within the Couchsurfing Web site; strangers do not normally share with each other their passions, life goals, life mission, and types of people they find interesting. Because the design of the Web site structures a response of relative self-disclosure online, the users involved in this interaction (looking at each other's profiles, gathering information about each other, interacting through email) become part of a dialogue that continues offline with an openness to self-disclosure and candid intimacy. Nana explained, "When I talk to people, I do not wish to discuss the superficial things like 'what went on in the football game' or 'which model has the biggest boobs,' so with Couchsurfer there's an excuse to avoid all that stuff because they're only there for a short time. So you get closer faster." But this feeling of "getting closer faster" is still instigated by the candidness found through the profile. Nana may approach her host in Stockholm by stating, "So you mentioned on your profile that you've been to Nepal. Tell me about that" or perhaps "I saw on your profile that you lived in South Africa and Italy. What were you doing there?" Reading another person's online profile provides an illusion that we in fact know the other person or, as stated earlier, we are familiar with them.

The amount of time Nana spends interacting face to face with her host is also linked to trust. From our analysis based on our network data, we have come to

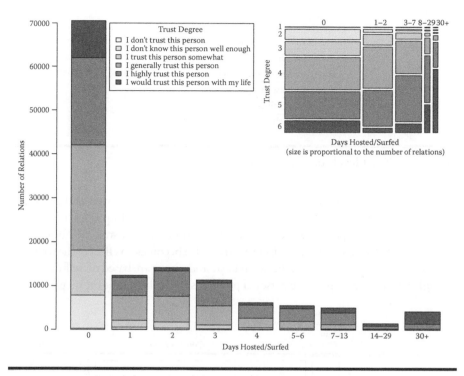

Figure 8.2 The dependence of trust levels on the time hosting and visiting.

conclude that those who hosted or visited one another, meaning those who spent a certain period of time with one another within a private space, are also more inclined to trust one another at a deeper level than those who had not hosted or visited and just exchanged emails online or met over coffee. Moreover, as we displayed in Figure 8.2, results from our data show that the longer the Couchsurfing visit, the stronger the trust is between individuals.

This is particularly worth noting if we want to analyze the context of a relationship as it relates to familiarization, and it forces us to question the link between the level of trust and the space in which interaction takes place. Within the hospitality exchange, trust develops at a high level, not contingent on a long period of time—the Couchsurfer usually stays for an average of only two days. The trustor (the host) has already allowed the guest into her controlled space; that is a large risk, bringing uncertainty and loss of control. Yet as the guest "proves" he is a good person, honoring the host's ownership and control of a space, trust between the two becomes quite strong, and this strength grows exponentially over time. Moreover, trust within this private space is reciprocated. The upper hand, or the power, lies with the person who has control of a space—in this case the owner of the home. Consequently, just as the host trusts the guest to act in a certain way within his controlled space, the guest trusts the host not to harm her in any way within this space. This is linked to what

	Percent
I don't know this person well enough	6.7
I don't trust this person	.4
I trust this person somewhat	11.5
I generally trust this person	34.3
I highly trust this person	34.7
I would trust this person with my life	12.3

Figure 8.3 Trust levels on Couchsurfing.

was shown earlier in Figure 8.2; the hosted or visited relations as well as the number of days individuals hosted or visited each other have a positive impact on the level of trust. The more two people hosted or visited the more they trusted each other.

We would like to underline the fact that trust also becomes an effect of the reciprocity of self-disclosure. Interactions between hosts and guests are based on self-disclosure through storytelling, explaining personal problems, fears, hopes, or future plans.

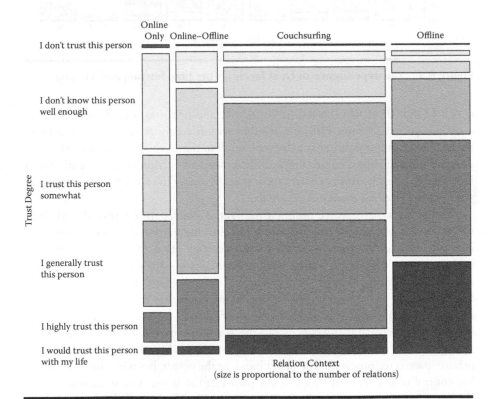

Figure 8.4 Trust levels among relations of a different origin.

Disclosing any sort of personal information involves psychological risk, such as the risk of rejection, risk of being judged, risk of being exploited, or the risk of being ridiculed. If despite this risk, one gains acceptance and openness from another person, then the other is more inclined to trust.

Moreover, if someone opens herself up before us, then we are more likely to trust her. If Nana's host in Stockholm tells her about her past personal problems, Nana would in turn be more inclined to trust her host because the host had risked something by speaking about her personal life. This large "gift" of self-disclosure is often reciprocal; Nana too would then feel inclined to talk about her life, creating a stronger bond of trust between her and her host. This reciprocity becomes a form of appreciation that creates a deeper sense of trust. Simon, a twenty-five-year-old from San Francisco, explained trust as a gift, stating that the more he risks in trusting the other person, the bigger the "gift" is and the more thankful the trusting party will be in receiving this gift. These risks are also involved in self-disclosure.

8.8 Inside the Private Sphere

The social navigation initiated online moves offline when the host and guest continue in interacting and exchanging trust. Without having a strict outlined space (e.g., a Web site) that offers certain affordances for appropriate interaction, the offline environment has less interactional structure and the users must socially navigate themselves through this new environment, making decisions and forming behavior. This creates, as Garfinkel (1963) outlined, a "perceivedly normal environment." Here, through interaction, the host and guest establish their conditions of play. The process can be outlined into stages that will explain the way in which the host and guest attempt to establish certain normative patterns of behavior in an offline setting.

First is the introduction stage, where the surfer and host meet in a public place or simply at the front door of the host's home and embrace or shake hands. Here, the surfer enters the private space of the host's home, the host shows the surfer his bed/floor space/mattress/etc. and gives him a tour of the house, and then they sit down to dinner or leave the house altogether in order to start touring the host's area. This is where the initial verbal exchange is initiated, as the individuals take turns giving a sort of monologue, building on their online profiles. After initial small talk, the first common biographical question we observed is centered on place, such as, "What brings you to Stockholm?" Or the guest may ask, "How long have you been living here?" Discussing physical place within this introductory dialogue is so common because both the host and guest are situated in the same apartment, in the same city. Yet, whereas the guest is a traveler, just passing through the given location, the host is a permanent resident and would know more about the place than the guest. This is why, instinctively, talk of place ends up being central to the introductory stage. This initial process of familiarization thus involves low risk and is engaged because it is based on a common subject among the two strangers.

The second stage, the insight stage, is the time in which one or both parties provide some insight into her own life, the lives of others around her, her personal history, her experiences, her problems, or her failures. As was stated earlier, these subjects tend to be the same subjects expressed on the online profiles. The one common theme in this stage is the presence of insight, which in turn raises the level of intensity of a conversation, thus raising the intimacy of the exchange. The insight stage can, but does not have to, include an exhibition of emotions, but it always includes a sense of trust for one or both parties. Eye contact and close, personal, spatial proximity is present. This process can last anywhere between an hour and several days, depending on how long the individual surfer stays with the host. It is worth noting that it is the insight stage that produces the close personal connection that individuals such as Nana longed for. It is in the insight stage that both the host and guest disclose intimate information about one another, heightening familiarity, which in turn, leads to a heightened degree of trust. Michael, a 30-year-old American, stated that on the "number of occasions" in which his hosts told him their life stories, he could tell that doing so was "exceptionally personal to them" and that they were "very involved in it." This discussion is for Michael, "part of what makes this whole experience seem so special. These things which come across as very emotional and very personal to the other person, just get given out as these gifts to the stranger that came by." As self-disclosure is looked at as a gift, often guests feel obliged to provide some travel stories, insight, or personal information about themselves in order to "give back" to the host. Both the host and guest attempt to create an environment where interaction seems "normal," despite the fact that they do not have a prior understanding of how to interact with one another. Michael explained that "you have this intimate relationship where you're sharing a physical space with someone that's very personal to you and so to equalize this physical intimacy you develop this emotional intimacy and that kind of makes people more comfortable. Because as soon as they get really close and really personal with someone they don't seem to feel that strange about how physically intimate the relationship is."

As strangers become physically close or are confined to a set area, their physical closeness is out of balance with the emotional closeness or intimacy level they have with each other, and it becomes simply natural to try to "catch up" with the physical intimacy and start talking about oneself in order to avoid the feeling of awkwardness. The feeling of comfort, as Michael explained, is crucial. Implicit rules of play are thus established in the insight stage that allow the host and guest to feel comfortable with one another. The more comfortable the interaction between the host and guest, the more trustworthy they will feel to one another.

The third stage, termed here the embedding stage, is where the surfer has to leave to go to her next destination, and both parties are faced with the decision of whether or not to keep, or embed, this new friend in their span of friends. This decision is based on the intensity of exchange during the insight stage and the amount of intimacy and insight experienced by both parties. Michael described many of his conversations as "special" or as a "gift." The larger the "gift" is, meaning the more the host/guest opened up to the other, the more likely the person will be embedded within their memory.

This process of embedding is not at all physical; it is not about scribbling a name down in an address book or adding an email address to one's email account. According to our online survey, Couchsurfers keep in touch with 50% of their hosts/surfers. Within this entire process, the introduction stage must occur in some shape or form but does not have any value for the embedding stage. If the host and guest felt comfortable in their interaction and a certain level of self-disclosure occurred (in the insight stage), then the two are more likely to embed one another in their list of friends. This "list" is both the online buddy list within the online social network and part of one's imagined social network—the process in which we accept a friend into our lives. When someone is in our imagined social network, a number of friendship practices are adopted that involve this given individual, including introducing her to other friends and communicating with her on a "regular basis."

8.9 Embedding Trust Online

Let's return to Nana and assume that she has interacted with her host and returns to her home in Helsinki. She will soon after register her experience online. Couchsurfing offers two affordances for her to do this (a) Nana can add her host in Stockholm to her friendship list, and (b) Nana can give her host a reference. Both will show up on Nana's profile, as will the level of trust.

Based on the online friendship links, the declared trust is high among Couchsurfers, as 47% of relations between users were declared as high trust relations. Only 0.4% of registered relations are distrust relations. In almost 7% of relations, users were not able to evaluate the level of their trust to another person.

The embedding stage leads us to a question: How has the process of familiarization online as well as the interaction offline led to a sense of trust between the host and guest? Furthermore, does it matter if users meet each other face to face, or is it possible for a virtual community to build trust strictly online?

In order to answer this question, we analyzed a large amount of anonymous whole network data provided by the administrators of Couchsurfing.com. Our data set was collected in the beginning of February 2006, when Couchsurfing had almost 45,000 active users,[3] including 221,180 friendship dyads, which included various variables, such as the origin and the duration of acquaintanceship as well as how strongly the individuals trust one another on a scale of 1 to 6.

Our findings helped us understand that the context of acquaintanceship influenced the perception of trust. Those relationships that were created in an offline setting show a much higher trust tendency than those of people who know each other through Couchsurfing. By far, the most trustworthy are those relationships in which people know each other from an offline setting.

Offline friendships are those of Couchsurfers who stated that they knew each other before joining Couchsurfing. Here, with 33% of individuals giving each other the highest possible trust degree, 41% state they would highly trust the other person. This is significantly different from the strictly online connections, where only 2.1% of people gave out the highest trust degree, and, comparatively, 11% would highly trust each other. These online connections are those who interacted through email or the Couchsurfing chat room.

Yet in the context of this study, what is also worth noting is the degree of trust found among those who have met online and then either hosted or visited another member of the community. It seems that this face-to-face interaction makes a great difference to the trust degree. Compared to the online-only group, 39% of those dyads who engaged in Couchsurfing would highly trust each other, and 7% gave each other the highest degree of trust. Thus, individuals who know each other from face-to-face interactions are more likely to trust each other than are those who only know each other from an online context. Moreover, when people meet online and then offline and risk something by explicitly engaging in a trust exchange (as in hosting/surfing), they are more likely to trust one another than are those who did not engage in this exchange (only online–offline.)

8.10 Conclusions

The high level of trust between Couchsurfers may be an effect of what Coleman once identified as purposive action in which "individuals sometimes have a strong need to place trust" (Coleman 1990). Coleman explained this phenomenon using an example of a girl who trusts a boy to walk her through the woods despite not knowing the boy. This girl is lonely and needs companionship, which is potentially available but only if she places trust in the boy. Coleman assumed that both actors in the trust exchange are purposive, having the aim of satisfying their interests, whatever those might be (Coleman 1990). Couchsurfers may have no choice but to place trust in one another in order to fulfill their purpose of finding accommodation, meeting other travelers, interacting with locals, learning about the world, or gaining access to a local's view of a place. This is true only if we disregard the fact that members of this virtual community make choices of whom to trust and not to trust. If Nana had been given the choice of only one host in Stockholm, she could potentially adopt a purposive trust that Coleman discussed. Yet as is often the case with Couchsurfing, trust is a choice. Moreover it is a choice among self-selected individuals who have a relatively similar approach to trust, strangers, and interest in others.

Users choose whom to trust based on the perceived level of familiarity they gain from another user. As we analyzed here, a virtual community offers affordances to members via their online profile that helps communicate personal information, creating a sense of familiarity between users. If a Couchsurfer makes use of these affordances and creates a detailed profile of himself, he is in turn creating a space in which anyone viewing his profile can become familiar with who he is. Often members of this virtual community do not use the affordances offered by Couchsurfing, meaning, they do not fill in some areas in their profile, they do not attach a picture, or perhaps they do not add a friend to their "friendship list." When another community member views this type of profile, she is less likely to gain a sense of familiarity from this user.

As we explored, familiarity does not always directly lead to trust. If Nana had a choice of two hosts in Stockholm, and both their profiles included a lot of personal description, Nana would have to choose whom she wanted to visit, and the decision would be based on whom she trusts more. This is where the idea of homophily provides us with some answers: We are more likely to trust those whom we know better, and often we are more likely to meet and interact with people who are similar to us. Even though Nana may feel familiar with both potential hosts in Stockholm, choosing the host who seems more similar to her will increase the likelihood of fluid interaction (having, for example, common topics to discuss.) Homophily also increases the chances of establishing a "perceivedly normal environment," as it is generally easier to predict how someone similar to yourself will act than it is to interact with someone whom you do not understand, do not relate to, or do not share a cultural background with.

It seems that the more choice there is within an online–offline virtual community, the more social navigation an individual has to do in order to come to the conclusion of whom to interact with and whom not to interact with, or in the case of Couchsurfing, whom to trust and not to trust. The concept of social navigation is crucial in understanding the way in which users come to certain decisions in this array of choice found in such a community. Specific design affordances that we discussed such as the recommendation system help the user socially navigate throughout the virtual space. From this point of view, when interaction moves form an online to an offline sphere, the process is not fragmented; information provided in the online profile allows individuals to meet each other face to face for the first time, already being familiar with one another.

References

Baudrillard, J. 1994. *Simulacra and Simulation*. Ann Arbor: University of Michigan Press.

Beck, U. 1992. *Risk Society: Towards a New Modernity*. London, UK: Sage.

Dieberger, A. 1997. Supporting social navigation on the World Wide Web. *International Journal of Human-Computers Studies* 46 (6):805–825.

Dieberger, A. 2003. Social connotations of space in the design for virtual communities and social navigation. In *Designing Information Spaces: The Social Navigation Approach*, edited by D. B. Kristina Höök and Alan J. Munro. London, UK: Springer.

Dieberger, A., P. Dourish, K. Hoeoek, P. Resnick, and A. Wexelblat. 2000. Social navigation: techniques for building more usable systems. *Interactions* 7 (6):36–45.

Ellison, N., R. Heino, and J. Gibbs. 2006. Managing impressions online: Self-presentation processes in the online dating environment. *Journal of Computer-Mediated Communication* 11 (2):415–441.

Gaver, W. W. 1992. The affordances of media spaces for collaboration. Computer supported cooperative work. *Proceedings of the ACM conference on Computer Supported Cooperative Work.* Toronto, Canada: ACM. p 17–24.

Gibson, J. J. 1979. *The Ecological Approach to Perception.* Boston: Houghton Mifflin.

Goffman, E. 1959. *The Presentation of Self in Everyday Life.* New York: Anchor.

Granovetter, M. 1973. The strength of weak ties. *The American Journal of Sociology* 78 (6):1360–1380.

Luhmann, N. 1979. *Trust and Power.* New York: Wiley.

Papacharissi, Z. 2002. The presentation of self in virtual life: Characteristics of personal home pages. *Journalism and Mass Communication Quarterly* 79 (3):643–660.

Sztompka, P. 1999. *Trust: A Sociological Theory.* Cambridge: Cambridge University Press.

Ziegler, C. N., and Golbeck, J. 2007. Investigating interactions of trust and interest similarity. *Decision Support Systems* 43 (2):460–475.

TYPES OF ONLINE SOCIAL ENVIRONMENTS

III

TYPES OF ONLINE SOCIAL ENVIRONMENTS

Chapter 9

Second Life™, Immersion, and Learning

Diane Carr and Martin Oliver

Contents

9.1 What Is *SL*?

Second Life™ (*SL*) (http://www.secondlife.com) is an online, graphically rendered, three-dimensional (3D) virtual world that is accessed with an "avatar" or online personification. Users, or "residents," explore and build neighborhoods, make and model fashions, build spaceships, design gardens, run businesses, exhibit pictures, meet others at conferences, and throw parties. As this indicates, there are many different ways to participate. *SL* is a toolset, a virtual world, and it supports social networks. It is a platform for various forms of play and a place where people gather to create or play games. Unlike a game, however, *SL* does not present the resident with a set of goals to be pursued according to rules in order to achieve a winning outcome (goals, rules, and win/lose states generally feature in definitions of "game" as distinct from "play"; see Salen and Zimmerman 2004). *SL* residents may not be confronted with a set of game rules, but there are constraints in place, including the developer's terms of service and social etiquette. Although there may not be a specified mission or goal, there are still goal-oriented practices and aspirations in the form of expertise, display, reputation, or popularity, for example.

After downloading *SL*, a new resident (newbie) is presented with a generic avatar. There is a convention that an unmodified avatar is a sign of inexperience (see Boostrom 2008). By going to an edit menu, a resident can access a set of sliding scales to alter their avatar's "shape" (height, arm length, bottom size, etc.) and "skin" (Figure 9.1). The desire to further individualize an avatar may lead to shopping expeditions, or a resident might prefer to take a "do-it-yourself" approach. For example, in the case of the avatar shown in Figure 9.2, Jelly, a translucent layer has been added to her face—a "skin" that was created using graphics software and a downloadable template. Make-up, freckles, and a "beauty spot" were painted on, and the resulting image was uploaded to *SL* and applied. Body shapes, like skins, can be modified. Residents seeking something more sophisticated or less humanoid (see Figure 9.3) than the default options might go shopping (see Figure 9.4) or create a shape for themselves using a 3D modeling application (guides and templates can be found here: http://secondlife.com/community/templates.php).

As this suggests, an avatar might be a composite of found, made, and purchased elements. In Figure 9.2, for example, Jelly is wearing hair purchased from a wig shop. The glasses are from a spectacles shop. Her gray shirt was purchased, but her

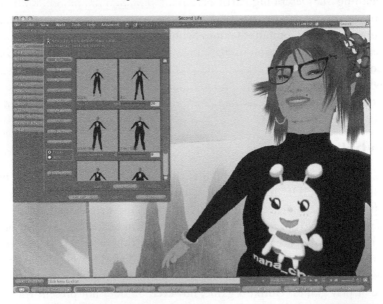

Figure 9.1   Editing appearance. (Used with permission of Linden Labs.)

Figure 9.2 Jelly wearing a human female "look." (Used with permission from Linden Labs.)

Figure 9.3 Jelly wearing a different look. (Used with permission from Linden Labs.)

Figure 9.4 Window shopping for a body. (Used with permission from Linden Labs.)

dark jacket, hoop earrings and eyes were all promotional "freebies." Freebies are objects or scripts given away by residents and businesses within *SL*. Thanks to freebies, an intrepid new resident can equip him- or herself for little or no outlay. Jelly's facial expression is the result of an animation: a piece of script (or programming) that makes her smile. Fortunately the smile comes and goes, rather than remaining fixed. She walks a certain way because a script attached to one of her shoes triggers a particular animation. Different walks ("sexy walk," "power walk") are available for purchase or from freebie warehouses.

Residents seeking something particular in a script—skin, body, outfit, vehicle, facility, or environment—have the option to build or create it for themselves. Learning the relevant skills may involve trial and error, accessing tutorials within *SL*, or following online guides created by *SL* experts. Although only "landowners" can create permanent installations, anyone can use open-access areas known as sandboxes to experiment on a large scale and meet fellow builders.

Whereas some residents choose to spend their time building, others prefer to meet for socializing, sex, or themed role-play. This variability is one of the reasons *SL* is difficult to define, and it is also one of the reasons *SL* can be disorientating or even alienating for new residents. The new player of an online game such as *World of Warcraft*™ (see Chapter 10) is introduced to the game and the interface through a set of tips and a series of missions of increasing difficulty and complexity. The game gives players instant feedback on the activities undertaken in the form of points. There is room for free exploration and playful experimentation, but the player is also provided with a clear series of goals. These kinds of guidelines are not present (or present in the same way) within *SL*, which new arrivals might experience as comparatively confusing or even pointless. It is interesting to think about these potential pleasures and frustrations in terms of affect, space, and navigation.

Janet Murray (1997) has described navigation in virtual spaces and the pleasures they evoke, using the model of the maze and the rhizome. The maze is a relatively directed experience that has the disadvantage of moving "the interactor toward a single solution [. The] desire for agency in digital environments makes us impatient when our options are so limited." Murray draws on the philosophy of Deleuze to propose the rhizome as a contrast to the maze. The rhizome, like a complex root system, allows unrestricted exploration. The disadvantage of the rhizome is that it may lead to disorientation. Thus, argues Murray, both "the overdetermined form of the single-path maze and the underdetermined form of [the rhizome] work against the interactor's pleasure in navigation" (p. 134).

Murray proposes the labyrinth as a version of navigation that incorporates the freedom of the rhizome with the purposefulness of the maze. The labyrinth is "goal driven enough to guide navigation but open-ended enough to allow free exploration" (p. 135). Our early outings in *SL* were given a minimal, labyrinthine structure by our desire to modify and personalize our avatars. This goal led to motivated exploration, and our experiments generated instant results. The sense that our avatars were "self-contained" and customizable offset, to some extent at least, the rhizomic disorientation we experienced on arrival.

9.2 *SL* and Education

SL is not, of course, the only virtual world available, nor is it the most popular (http://www.mmogchart.com/), yet it is the virtual world that educators have focused on recently, apparently because it offers "a relatively stable, relatively accessible, inexpensive and inhabited, persistent world where it is possible to build simulations, labs or locations" (Carr 2008a). These potentials are reflected in the variety of approaches currently taken by educators working in *SL*. Some educators focus on technical aspects (running classes on scripting, for example), fostering collaboration, or designing educational simulations. Others might teach specific curriculum content through discussion, demonstration, role-play, or practical exercise. Some might approach *SL* and its community as phenomena, as something to learn about, rather than simply a venue for learning, or a tool for teaching. In other contexts, or at other times, educators might simply regard the virtual world as a convenient location for a class meeting.

Within *SL* there are numerous examples of sites, buildings, installations, and simulations that are intended to serve a teaching, learning, or research agenda. One example would be the Social Simulation Research Lab,[1] which hosts a collection of materials on Internet research and social networking theory. There are museums dedicated to arts or particular sciences, as well as didactic models and simulations that address issues of health or anatomy and various themed areas designed for language learning. Virtual versions of real-world colleges and universities are scattered throughout *SL*.

Of course, learning is not limited to classrooms, and curriculum attainment is only one indication of learning. Alternative notions, for example, that learning is a social achievement and is evidenced through competent activity within a community (see Wenger 1998) are also relevant. Communities form around particular sites (such as sandboxes) and practices (such as machinima.) People learn by asking each other questions. Residents create expert tutorials and design facilities to teach their fellow residents *SL*-specific skills. A well-known example would be The Ivory Tower of the Primitives, which avatars can explore while accessing note cards, exercises, and tutorials in order to learn how to build in *SL*.

Those researching education and learning in *SL* come from a range of disciplinary backgrounds and, as a result, they employ a variety of research strategies. Methods include forms of textual analysis, observation, interview, surveys, and tests of various descriptions. These later, more positivist styles of enquiry might be

---

[1] In addition to the Social Simulation Research Lab (Hyperborea 218, 90, 24), well known examples of educational sites would include the International Space Flight Museum (SpacePort Alpha 47, 82, 23), and Virtual Hallucinations (Sedig 26, 43, 21). For a list of places to visit, see http://secondlife.iste.wikispaces.net/page/code/*SL*tours. Accessed February 2009.

instantiated within *SL* through the development and use of context-specific data collection methods, such as survey booths. Interaction logs or recordings of behavior in specific settings may also be used as evidence. All research involves considerations of ethics and ethical practice, and this is obviously of particular concern when humans (via their avatars) are asked to participate in experiments, not least because the situations they encounter are the direct responsibility of the researchers who design them. There are several determinants of what currently counts as ethically acceptable research in *SL*, including the terms of service specified by Linden Labs (*SL's* developers) as well as more conventional points of reference, such as the codes of conduct of relevant professional organizations and the guidelines that have been produced by various associations (such as Ess and the AOIR 2002).

Another strand of research into education in *SL* looks to issues of access and inclusion (Sheehy 2008). *SL* is not accessible to everyone, for reasons of geography and broadband, hardware requirements, or disability. Some disabled people, however, have found it empowering (Boellstorf 2008 p. 147), and residents have created resources to facilitate participation (see more information at www.VirtualAbility. org). In addition to issues of usability and access, *SL* allows researchers to examine the social dimensions of disability. During 2007, for example, an integrated voice feature was added to *SL*. Until this point the default mode of conversation within *SL* had been text-based chat ("talking" by typing.) The new feature raised the possibility that voice (the spoken voice of the user input through a microphone, heard through headphones) would become the "normal" way to communicate in-world. This new feature was assumed by some to be an asset for education and commerce, but protests were lodged and concerns were raised by various groups, including residents who identified as deaf (Carr et al. 2008). Dismayed deaf people posting to *SL* forums found themselves on the receiving messages of support, as well as hostility and accusations of selfishness. As such tensions make apparent, disability can be considered a social as well as a technical issue within virtual worlds.

9.3 Defining *SL*, Describing Participation

The introduction of the integrated voice feature to *SL* during 2007 was controversial. At the heart of these tensions was a debate over who it is that *SL* is "really for" and what *SL* is, or should be. At times the discussion was framed in terms of different types of participation. The impact of these debates on *SL's* community has been noted. As part of our recent research (see "Acknowledgments") into learning in *SL*, for instance, we interviewed researcher Greg Wadley about his investigation of voice in *SL*, and the manner in which the new feature was being received by *SL* residents:

Greg: I am detecting two distinct groups of *SL* users. Some of the "old school" users (I don't really have a good name for them as a group) live in *SL* as an alternative fictional world. But *SL* is being taken up by a newer group—business and education people primarily—who treat it quite differently. Some people call these two groups "immersionists" and "augmentationists." Overgeneralizing, the former dislike voice while the latter welcome it.

It appears that these terms, "immersionist" and "augmentationist," were proposed by Henrick Bennetsen (Bennetsen 2006). Elsewhere the classifications have been elaborated on by *SL* commentators, including Akela Talamasca, who wrote that

> The Immersionists believe that *SL* is a complete and discrete world in itself, and should have no truck with anything to do with RL [real life]. One could gather roleplayers [...] into this camp. Immersionists tend not to disclose any of their RL information [...] The Augmentationists view *SL* as an extension of their RL, more as a tool to be used to interact with others. These residents see nothing wrong (in general) with more interaction and connectivity with RL. (Talamasca 2006)

Yet, for other commentators, this differentiation is problematic and inaccurate. Tateru Nino, for example, has asserted that:

> The whole seeming-struggle between augmentationists and immersionists is just so much smoke. There aren't two camps going on here. They're not duelling to death over the future of the platform. It's mostly a false *Us* and *Them* distinction that distracts and diverts quite unnecessarily. (Nino 2007)[2]

As educators interested in *SL*, what strikes us as important in the "augmentationist versus immersionist" debate is not the defining of particular kinds of *SL* users as much as the evaluative tone of such categorizations. The debates offer us an opportunity to look at conflicts over what constitutes legitimate and credible participation in *SL*. These debates are associated with a technical feature—the introduction of the integrated voice feature—which was regarded by some as an appeal to education and business sectors at the expense of "real residents."[3] So, it is not the definitions per se, that are interesting but the way that people choose to use these terms and the evaluative connotations that these terms accrue through this use. It is possible to extrapolate and imagine the strategic

---

[2] Nino's comments recall the "ludology verses narratology" debate within game studies. This debate has been important in terms of the emergence of the field, yet it suffers from similar problems of specificity and definition.

[3] For more on this issue, see "World or Platform: The Reality of *SL*" an in-world meeting reported at http://www.tikomatic.com/blog/2006/10/immersion_versu.html. Note how the title of the meeting recalls Richard Bartle's comments about the use of integrated voice features in virtual worlds quoted elsewhere in this chapter. See also "The big controversies in *Second Life*," written in 2006 by Gwyneth Llewelyn, and available online at http://gwynethllewelyn.net/2006/09/17/the-big-controversies-in-second-life.

employment of such connotations. What if, for instance, an educator describing him- or herself as an "immersionist" in particular contexts was not just a statement of preference but a way of staking a claim to legitimacy-by-association with the "real" *SL* community; with the progressive, the utopian, the credible, and the alternative?

This debate may be "old news" in the *SL* community, and the terms "augmentationist" and "immersionist" do not appear to have had much of an impact within *SL* and education circles. However, the terms "immersion" and "immersive" appear very frequently in conferences and journal articles on topics pertaining to virtual world pedagogy. Again, it is not just definitions that interest us. Rather, it is the idea that a term relating to a form of participation in *SL* can be vaguely defined and yet accrue evaluative or positive connotations. For these reasons we believe that it would be productive to take a closer, cautionary look at the terms "immersion" and "immersive." Theoretical models of immersion drawn from computer games studies make this closer examination possible.

9.4 Immersion

By our count the terms "immersion" and "immersive" appear approximately 15 times in the "2008 Autumn 'Snapshot' Report on UK Higher Education Research in *Second Life*" prepared for the Eduserv Foundation by John Kirriemuir (2008), more than 40 times in the 2007 proceedings from the *SL* Community Conference's Education Track, and approximately 180 times in the proceedings from the Open University's Conference ReLive08: Researching Learning in Virtual Environments (where contributors focused almost exclusively on *SL*.) In many cases, the term is used simply to mean a 3D environment—as in the phrase "immersive worlds"—and we have no argument with the term being used in this vernacular or general sense.

What is of greater concern would be instances where immersion remains only vaguely defined *and yet* is assumed to be "good" or "good for learning" or where it is only broadly or generally defined and yet approached as something that can be measured—as a phenomenon that will be supported or undermined by particular input devices or interface designs rather than others, as a property of the software, or as a form of participation associated with particular styles of pedagogy or learning. Clearly, in the context of such research, a vernacular definition is inadequate.

Clarity matters, because immersion is more complex and more paradoxical than is being supposed. For the sake of argument, let us begin by noting that immersion is associated with presence, with a sense of "being in the virtual world" (Lombard and Ditton 1997). Based on this, a hypothetical research team might decide that immersion correlates to some degree or in some way with realism, and follow this by asserting that realistic and detailed 3D worlds where avatars with photorealistic faces communicate while lip-synched to a user's real-time voice will be "more immersive." Although recognizable features or landmarks in a virtual campus might orientate new students (Carr 2008b), anyone equating increased realism with increased immersion would do well to consider Masahiro Mori's work on androids and the "uncanny valley." Mori pointed out decades ago that "as robots appear more humanlike, our sense of their familiarity increases until we come to a valley" (Mori 1970). If a robot is "too lifelike" it moves from being familiar to being disconcerting, strange, or scary. Recent game technology has borne out Mori's theory. Players and game critics have noted that insistently "life-like" avatars can be creepy or downright frightening (Thompson 2005).

When considering realism and detail, game and simulation researchers have noticed the value of a "less is more" approach. It has been proposed that selective exclusion has its benefits (Carr 2004). Swartout and Lindheim (2002), for example, explain that exact replication in military simulations (of a particular machine or cockpit, for instance) can be counterproductive. They argue that the more selective simulations favored by the entertainment industry are actually more effective in pedagogical terms. Game theorist Gonzalo Frasca makes a similar point regarding simulations, when he notes that Will Wright, the creator of *Sim City*, was influenced by Scott McCloud's work on comic books, "particularly the section when McCloud explains how the reader fills in the gaps of what happens between each panel of the illustrated story" (Frasca 2001).

Computer games analysis has shown that although various forms of design cohesion are undoubtedly important to satisfactory participation in games and online worlds (see McMahan 2003; King and Krzywinksa 2006), it does not follow that more realism or detail will lead to a greater level of immersion. Consider also these comments concerning realism, immersion, and the use of voice in virtual worlds made by Richard Bartle in 2003:

> If you introduce reality into a virtual world, it's no longer a virtual world: it's just an adjunct to the real world. It ceases to be a place, and reverts to being a medium. Immersion is enhanced by closeness to reality, but thwarted by isomorphism with it: the act of will required to suspend disbelief is what sustains a player's drive to *be*, but it disappears when there is no disbelief required. Adding reality to a virtual world robs it of what makes it compelling—it takes away that which is different between virtual worlds and the real world: the fact that they are *not* the real world.

This is not the first context where the term "immersion" has proven problematic. It is probable that the term arrived in *SL* and education circles from virtual reality (VR) research. As Sherman and Craig explain in *Understanding Virtual Reality* (2003), the term has been vaguely or even contradictorily defined within VR studies.

> The state of being mentally immersed is often referred to as having "a sense of presence" within an environment. Unfortunately, there is not yet a common understanding of precisely what each of these terms mean, how they relate to one another, or how to differentiate between them. (We have found one book in which chapters written by different authors give exactly the opposite definitions for immersions and presence.) (p 9)

In the discussion thus far, we have employed a rather skeletal definition of "immersion" or "immersive" in that we have discussed it in relation to a user being in some sense drawn into (or ejected from) a virtual environment. It is possible to consider the term and its applicability to *SL* and education in greater depth by referencing computer games scholarship in more detail.

9.5 Immersion and Computer Games Scholarship

Within computer games studies, considerations of the term "immersion" often begin with a reference to Janet Murray's (1997) *Hamlet on the Holodeck*, in which she proposes that

> The experience of being transported to an elaborately simulated place is pleasurable in itself [...] We refer to this experience as immersion. Immersion is a metaphorical term derived from the physical experience of being submerged in water. (p. 98)

In later games analysis, the notion of immersion is discussed in relation to a more critical form of attention, that of *engagement*. Drawing on VR studies, Alison McMahan (2003) has used Lombard and Ditton's (1997) review of the literature on presence to propose that immersion and engagement be attributed to different aspects of computer games. Immersion, in McMahan's analysis, is associated with the fantasy and narrative aspects of a setting, whereas engagement is linked to the challenges posed by the game play.

In other accounts (Douglas and Hargadon 2000; Carr 2006), engagement and immersion are not linked to particular aspects of a game or a text. Instead they are imagined as attentive states along a continuum that suggest a particular stance toward the game at a given moment. Immersion is experienced when a player is simply absorbed, and it is countered when the player is presented with a challenge that requires a more thoughtful "stepping back." This stepping back involves a more consciously critical or searching form of attention: engagement. The point is not that one mode of attention is "better" than the other or that the two are mutually exclusive. Instead, an engrossing gaming experience will involve constant attentive shifts. According to Carr (2006), players

> slip between immersed, close absorption and an engaged, critical distance. Thus it would not make sense to value engagement over immersion [...] the two states are complementary. The game is compelling, not because of its capacity to evoke either immersion or engagement [...] but because it allows the player to constantly move between the two. (p. 55)

It is easy to imagine how these shifting states of attention might be present in the processes of avatar customization that we outlined at the beginning of this chapter. Toying with the slider to manipulate the avatar's facial features or body type might be highly absorbing. The limitations of such manipulations might result in frustration that motivates the user to access alternative information or resources. These resources might in turn prove compelling. For example, a *SL* user might download a template and use graphics software to create more pleasing facial details for an avatar. These shifts "in" and "out" of *SL* and between less and more conscious forms of participation would be set off by various triggers. These triggers would alter in line with the user's level of expertise, because a task that is initially engaging might become a more immersive pleasure once the user attains competence.

In *Rules of Play*, game theorists and designers Katie Salen and Eric Zimmerman discuss "the immersive fallacy," defining the idea in this way:

> The pleasure of a media experience lies in its ability to sensually transport the participant into an illusory, simulated reality. According to the immersive fallacy, this reality is so complete that ideally the frame falls away so that the player truly believes that he or she is part of an imagined world. (p. 451)

Salen and Zimmerman take a strong position on this topic because they consider that game designers too often take the desirability of immersion and the association of immersion with photorealism for granted, and innovation suffers as a result. Salen and Zimmerman argue that falling prey to the immersive fallacy entails ignoring the dual nature or double consciousness of play. The player manipulates an avatar in a game world, but at the same time, the player is pursing game goals while considering various obstacles and the rules of the environment. Thus "the player is fully aware of the character as an artificial construct [and this] double consciousness is what makes character-based game play such a rich and multi-layered experience" (p. 453).

Obviously we should query the direct applicability of this argument to *SL* because of the generic and structural differences between this virtual world and games. In terms of unexamined propositions and "fuzzy" definitions, however, there is much here that is pertinent to current debates within the *SL* and education community.

We would add that it is certainly not the case that immersion has been consistently regarded as a neutral or positive phenomenon. At times, and in various contexts, the term has been associated with noncritical or passive pleasures that have been linked with reactionary ideologies. Consider, for instance, the influential twentieth-century critiques of popular music, conventional theater, and mainstream, narrative cinema, and the subsequent championing of concepts such as estrangement, reflexivity, and alienation (see Stam 2000, pp. 140–153, for a discussion of relevant theorists, including Adorno, Brecht, Wollen, and MacCabe. See also Ryan 1994; Dovey and Kennedy 2006).

In a similar vein, Gonzalo Frasca's work on computer and video games draws on drama theory to distinguish conceptually between immersive, passive pleasures, and critically conscious, active participation. In "Videogames of the oppressed:

Critical thinking, education, tolerance and other trivial issues," Frasca (2004) draws on the writing of Augusto Boal, particularly *Theater of the Oppressed* (Boal 1993) which was in turn influenced by Freire's (2000) *Pedagogy of the Oppressed* and Brecht's theories of drama) to propose a particular approach to the design of politicized games—one that involves the player moving into an active and critical position in relation to a simulation's design.

This, then, is another issue that educators speaking of the immersive power of *SL* have tended to omit from the debate: the suggestion from older theory that immersion is not inherently transformative. At times it has been regarded as the "opposite" of learning. According to such arguments, immersion would only be useful in pedagogical terms when it is combined with interruptions of various kinds that are designed to facilitate critique. This recalls Kurt Squire's (2002) comments about the use of commercial games in classrooms and the need for teachers to supply a critical framework.

From the discussion so far, it should be clear that different takes on the notion of immersion (and different definitions of the term) are present in computer games literature. Clearly, immersion is complex. We are not interested in asserting that there is one "right" way to conceptualize immersion, but we would propose that attempts to prove, facilitate, or measure immersion must be clear about what it is that is under investigation.

Two points have now been raised: First, that immersion is one form of attention, and that it exists alongside alternative modes of attention (such as engagement.) Second, that players simultaneously operate from two perspectives: They are "in the world" and yet outside it. If this is the case with players, we suggest that it would also be the case with students accessing *SL* within a particular course in a formal learning context. Together these points indicate that any theory of pedagogy in virtual worlds that links learning to immersion as if the two existed in a self-evident, mutually affirming manner or that conceptualized immersion as a sort of monotonous, monodirectional state, might not survive close inspection.

The various theories of immersion referred to above offer a useful starting point from which to explore the complexity of this concept. We do not want to argue that immersion is equivalent to audience passivity, and thus "bad," whereas other modes of participation are active, and hence "good." Rather, we would propose that experiences in *SL* involve various forms of attention and pleasure and various kinds of affect, which are triggered by a range of factors, depending on the user, on their level of expertise, and on the context of use. To explore this issue, we will briefly refer to our teaching in *SL*.

9.6 Teaching and Immersion in *SL*

The two taught sessions here under discussion took place during a module called "Computer-Mediated Communication" within a master's-level course an Information and Communication Technology in Education at the Institute of Education, University of London.[4] The course was taught via distance learning, and the students are adults and often teachers themselves. As noted, Salen and Zimmerman have argued that playing a game involves operating with a kind of "dual consciousness." The feedback that we got from our students was very positive, yet it suggested that (regardless of how "immersive" *SL* is) they experienced similar shifts in perspective and moved between "internal" and "external" frames of reference.

Consider "role" for example. Despite attending class in the guise of an avatar, feelings about the roles of student and tutor were carried into *SL*, and performance in *SL* was regarded as having real-life ramifications, as this student's comments suggest:

> [...] What was playing on my mind on that session was the tutor's perception toward the students' participation. Did they give credit to students who answered most of the questions [?...] How about a few students who sometimes give long answers and short answers? How about if one of the students had problems with the computer, like *Second Life* crashed and it took a long time sometimes to restart? The tutor might think that this student participates less and [is] not very active in the session.

What is notable here is that in this context, *SL* was not necessarily experienced as a "place apart," where role exists to be played with or subverted, or risks taken with no repercussions. Clearly, users bring their existing expectations to class with them. We did, however, see indications that real-time, virtual space contact with fellow students felt more "real" than other online contact to that point. In the following comments, for example, a student distinguishes between the "real class" and the rest of the course, which had been taught via distance learning and Blackboard (a more conventional virtual learning environment than *SL*, where communication is primarily through forum-styled exchanges.) For these distance learners, the *SL* sessions were felt to be particularly compelling for social reasons.

> I could feel the "real class" when I saw [a] bunch of you gathering at the outside of the ground floor. I felt that finally I would meet all my classmates (even though it was not real.) Can you imagine in real life when you meet your classmates for the first time and you will automatically introduce and ask around about people?

The students had in some sense already "met" one another via online discussions and posts on Blackboard before the *SL* sessions. Thus a particular set of relationships was already in play prior to their meeting in the virtual world. For most of the students, then, the sessions succeeded not because the virtual world itself was intrinsically engrossing but because of what it offered in contrast to the more

---

[4] For a much longer account of these sessions and student feedback, see "Learning to teach in *Second Life*," an online report (Carr 2008b).

conventional learning platform, in the context of a specific course, within a set of relationships, and while performing particular roles.

For us, this suggests that it would be a mistake for educators to invest too heavily in the notion of immersion for its own sake. Instead, it would be more productive to consider learning in *SL* in terms of multiple forms of participation and a range of possible attentive states, which may be more or less suited to specific learning contexts and which may be facilitated and supported by different pedagogies and practices.

Our students were "newbies," and this may be one reason they tended to make sense of proceedings according to external references, including the education context, the institution in which they were enrolled, our roles, and eventual assessment. There may be various ways to argue that all these factors can be circumvented or designed to be "more immersive" by becoming more integrated into *SL*; but it is not immediately clear what that would achieve, educationally. Our experiences suggest that it would be problematic to assume that "more immersed" is "better" for learning or teaching, especially if such an approach is not sensitive to the students' expectations, contexts, and interpretive frameworks.

9.7 Conclusion

In this chapter we have introduced the virtual world *SL* and explored some aspects of the debates concerning *SL* and education. This has involved tracking some curious twists in relation to the currency of a particular term: immersion. While the term has positive association within the *SL* and education community, older literature has proposed that immersion can be understood as an uncritical form of participation that facilitates the transmission of dominant ideologies by the powers-that-be. We have no wish to underestimate, demonize, or dismiss the pleasures of immersion, but we would query the pervasive and often uncritical use of this term within the *SL* and education communities.

There are various ways to experience, learn in, or learn about *SL*, and thus there can be no single "best practice" model for educators. We have mentioned the variability of *SL* and noted that there are different ways it might be used, enjoyed, or defined. *SL* can, in fact, be a disorienting, ambiguous place, especially for new users (Carr et al. 2008). This ambiguity should not be regarded as an obstacle to be overcome in order for learning to be successful. Indeed, although *SL* can be used as a place to build virtual seminar rooms or otherwise replicate elements of real-world education, it is also an opportunity to upset the roles of both educator and student and to expose the conventions that inform these roles in the process.

When theorizing virtual world pedagogy, it is important to appreciate that various forms of participation, attention, and affect may be part of the learner experience. Education in *SL* involves multiple frames of reference—personal, social, and technical—each of which may have implications for learning. To further our understanding of the pedagogic potentials of *SL*, we need to be clear about which of these frames we draw on when making particular claims and be specific about the concepts that we employ in our research.

Acknowledgments

Elements of this research were undertaken during 2007-2008 with the support of the Eduserv Foundation. Full details of this research project, which was titled "Learning from online worlds: Teaching in *Second Life*" can be found at the project blog: http://learningfromsocialworlds.wordpress.com/.

References

Bartle, R. (2003) Not yet, you fools! At game girl advance. Online at http://www.gamegirladvance.com/archives/2003/07/28/not_yet_you_fools.html. accessed November 2008.

Bennetsen, H (2006) Augmentation vs immersion. Online at http://slcreativity.org/wiki/index.php?title=Augmentation_vs_Immersion, accessed November 2008.

Boal, A. (1993) *Theater of the Oppressed*. Communications Group Theatre.

Boellstorf, T. (2008) *Coming of Age in Second Life: An Anthropologist Explores the Virtually Human*. Princeton, NJ: Princeton University Press.

Bootstrom, R. E. (2008) The social construction of virtual reality and the stigmatized identity of the newbie. *The Journal of Virtual Worlds Research* 1(2), November. Online at http://www.jvwresearch.org/v1n2.html, accessed February 2009.

Carr, D. (2004) Modelled cities, model citizens; from overseer to occupant. In *Sim City: the Virtual City*, ed. Bittanti, M. Milano, Italy: Edizioni Unicopli. pp. 193–210

Carr, D. (2006) Play and pleasure. In *Computer Games: Text, Narrative and Play*, eds. Carr, D., Buckingham, D., Burn, A., and Schott, G. Cambridge: Polity. pp 45–58.

Carr, D. (2008a) Learning in virtual worlds. In *Education 2.0? A Commentary by the Technology Enhanced Learning Phase of the Teaching and Learning Research Programme*, ed. Selwyn, N. TLRP, pp 17–22.

Carr, D. (2008b) "Learning to teach in *Second Life*." Report for Learning from Online Worlds; Teaching in *Second Life*. London: Institute of Education/Eduserv Foundation, Online at http://learningfromsocialworlds.wordpress.com/learning-to-teach-in-second-life/, accessed February 2009.

Carr, D., Oliver, M., and Burn, A. (2008) Learning, teaching and ambiguity in virtual worlds. Paper presented at ReLive08 at the Open University, Milton Keynes, UK, November 2008. Online at http://learningfromsocialworlds.wordpress.com/paper-for-relive-08-at-the-ou/, accessed February 2009.

Douglas, J. Y., and Hargadon, A. (2000) The pleasures of immersion and engagement: Schemas, scripts and the fifth business. *Digital Creativity* 12: 153–166.

Dovey, J. and Kennedy, H. W. (2006) *Game Cultures* Milton Keynes. Open University Press.

Ess, C., and the Association of Internet Researchers [AOIR]. (2002) Ethical decision-making and Internet research: Recommendations from the AOIR Ethics Working Committee1. Online at http://aoir.org/reports/ethics.pdf, accessed February 2009.

Frasca, G. (2001). The Sims: Grandmothers are coolen than trolls. in *Game Studies* 1(1), July 2001. Available at http://www.gamesstudies.org. Accessed August 15, 2009.

Frasca, G. (2004) Videogames of the oppressed: Critical thinking, education, tolerance and other trivial issues. In *First Person: New Media as Story, Performance, and Game*. eds., Wardrip-Fruin, N. and Harrigan, P. Cambridge, MA. MIT Press.

Freire, P. (2000) *Pedagogy of the Oppressed*. Continuum. pp. in 85–88.

Galarneau, L. (2005) Spontaneous communities of learning: Learning ecosystems in massively multiplayer online gaming environments. Proceedings of the DiGRA 2005 conference: Changing Views—Worlds in Play, Vancouver, Canada. Online at http://www.digra.org/dl/db/06278.10422.pdf, accessed September 2008.

King, G., and Krzywinska, T. (2006) *Tomb Raiders and Space Invaders*. London, UK: I.B. Tauris.

Kirriemuir, J. (2008) The 2008 autumn "snapshot" report on UK higher education research in *Second Life*. Prepared for the Eduserv Foundation, Bath, UK. Online at http://www.eduserv.org.uk/foundation/sl/uksnapshot102008, accessed February 2009.

Lombard, M., and Ditton, T. (1997) At the heart of it all: The concept of presence. *Journal of Computer-Mediated Communication* 3(2), September. Online at http://jcmc.indiana.edu/vol3/issue2/lombard.html, accessed February 2009.

Masahiro, M. (1970) The uncanny valley. *Energy* 7(4):33–35. Translated by Karl F. MacDorman and Takashi Minato. Online at http://www.androidscience.com/theuncannyvalley/proceedings2005/uncannyvalley.html, accessed February 2009.

McMahan, A. (2003) Immersion, engagement and presence: A method for analyzing 3-D video games. In eds. Wolf, M., and Perron, B. *The Video Game Theory Reader*. New York: Routledge. pp 67–86.

Murray, J. H. (1997) *Hamlet on the Holodeck*. Cambridge, MA: The MIT Press.

Nino, T. (2007) Immersion versus augmentation. August 24, 2007. *Second Life* Insider. Online at http://www.secondlifeinsider.com/2007/08/24/immersion-versus-augmentation/, accessed November 2008.

Oliver, M., and Carr, D. (in press) Learning in virtual worlds: Using communities of practice to explain how people learn from play. *British Journal and Education and Technology* 40(3).

Ryan (1994) Immersion vs. interactivity: Virtual reality and literary theory. *Postmodern Culture* 5(1): pp 447–457.

Salen, K., and Zimmerman, E. (2004) *Rules of Play*. Cambridge, MA: The MIT Press.

Sheehy, K. (2008) Virtual Environments: Issues and opportunities for developing inclusive educational practice. Paper presented at ReLive08 at the Open University, Milton Keynes, UK, November 20–21, 2008.

Sherman, W. R., and Craig, A. (2003) *Understanding Virtual Reality: Interface, Application and Design*. San Francisco, CA: Morgan Kaufmann Publishers.

Squire, K. (2002) Cultural framing of computer/video games. *Game Studies*, 2(1). Online at http://www.gamestudies.org/0102/squire/, accessed February 2009.

Stam, R. (2000) *Film Theory: An Introduction*. Oxford, UK. Blackwell Publishing.

Swartout, W. and Lindheim, R. (2002) Does Simulation need a reality check? SimScience Workshop, Scientific Exploration of Simulation Phenomena, at the Simulation and Warrgaming Centre, National Defense University, Washington, DC. June 6–9 2002. PDF online at http://www.sim-summit.org/simsummit/0309/03090.pdf accessed July 2009.

Talamasca, A. (2006) Immersionist or augmentationist? September 25, 2006. *Second Life Insider.* Online at http://www.secondlifeinsider.com/2006/09/25/immersionist-or-augmentationist/, accessed November 2008.

Thompson, C. (2005) Monsters of photorealism. Online at http://www.wired.com/gaming/gamingreviews/commentary/games/2005/12/69739, accessed November 2008.

Wenger, E. (1998) *Communities of Practice: Learning Meaning and Identity.* Cambridge, UK: Cambridge University Press.

Chapter 10

Residency, Relationships, and Responsibility

The Social Side of World of Warcraft™

David White

Contents

Focusing on the popular *World of Warcraft*™ (*WoW*) game, this chapter explores the ability of massively multiplayer online games (MMOs) to support not simply a disparate collection of gamers but also a society of players. Drawing on both qualitative ethnographic studies and quantitative data produced by the game, the chapter explores the journey long-term players make as they mature from individual "newbies" to fully fledged community members, examining the tension between "game play" and social motivations as they progress through the MMO.

By stripping away the genre of the MMO, it is possible to construct an understanding of this new platform for society—not in the context of sword-wielding dragon slaying but in terms of kudos, social capital, and community. The often used moniker "World of Chatcraft" is indicative of the MMO as a sociable space that is fundamentally distinct from traditional console gaming. The MMO can become a place to form relationships facilitated by a shared understanding of the game world. This world does not sit apart from "real life" (RL) but intersects with it, allowing players to live out a portion of their lives online.

The MMO has a rich heritage that can be traced from board games, through text-based digital forms, to the increasingly lush, visually immersive environments. The genre's earliest appearance on a computer network can be traced back to the Multi-User Dungeon (MUD) created by two developers at Essex University that spawned a genre of networked games. The MUD simply described the environment, options, and interactions in text form, allowing the player to explore, fight monsters, and communicate with nonplayer characters using simple commands such as "Go North," "Use Potion," and "Ask for Key." Crucially these games also allowed players to talk to each other remotely using local computing networks and then increasingly the Internet. The lack of visuals was an advantage for these early multiplayer games, as the individual's imagination realized the environment and was not alienated by unsatisfactory renderings of castles and dungeons.

The fans of MUDs tended to be a subculture of computing specialists and fantasy genre enthusiasts, as the technology was not readily available and the style of interaction was akin to a form of programming. However, we can look to the MUD as the first "online" space in which it was possible to progress through the structures of a game while potentially socializing with fellow players at a distance. These text-based MMOs, which now appear rudimentary next to their contemporary visual counterparts, acted as a hub for relationships and communities and are a clear precursor to the culture that suffuses the sound and fury of Blizzard Entertainment's *WoW*.

It is important to remember that beneath the behemoth of *WoW* lurks an ambivalent database, which is a slave to the mathematics and command line style instructions that arose from dice-throwing board games and text-based MUDs. However, it was the audiovisual feast of *WoW*, not the game mechanism that drove the genre beyond a core audience of fanatics and enticed a broad spectrum of players into its world. The aesthetic crossed a psychological threshold by providing not the technologically unwieldy "virtual reality," but what has been termed "good enough reality" that allowed players to feel immersed in the fictional world of Azeroth. The genre no longer relied so heavily on player engagement based on enthusiasm for the fantasy milieu. It could now entice individuals with a form of engagement that shares many factors with cinema. The filmic qualities that bring many players to *WoW* are increasingly supplemented by interpersonal opportunities as the player progresses through the game.

10.1 Scale and Statistics

As of autumn 2008 *WoW* has over 10,000,000 subscribers (Woodcock 2008), each spending on average somewhere between ten and twenty hours a week in-world. The average age of *WoW* players is twenty-six, with only approximately 25% of players being teenagers (Yee 2006). The game has an internal economy of "play money" that, despite being against Blizzard's terms of use, is traded on the Internet in a similar manner to RL currencies and has a specific dollar value. Since its launch in 2004, subscriptions to the game have steadily increased with, as of autumn 2008, no signs of leveling off. This popularity cannot be accounted for by "game play" alone and would appear to be, in part, the shared experience of being in an online environment with others. The relationship between the game as technology and the game as a culture is complex and presents a challenge to the researcher attempting to gain an understanding of what motivates players to engage and why communities form in and around *WoW*.

10.2 Research Methodologies

In *WoW* there are many modes of communication, including group text chat, private messaging, and voice chat plus a wide range of locations and activities, each with their own contexts and implications. Capturing the activity of a group of players presents similar problems in research terms to studying any RL collection of individuals who could be loosely described as a community. To come to a valid understanding of how the players and the game function as a system, a balance needs to be struck between the highly qualitative immersion of the ethnographer and the quantitative approach of the statistician.

10.2.1 Reflective Observation and Interview Technique: Hybrid Virtual and RL Ethnography

The later stages of the game are the most complex and collaborative, involving multiplayer groups formed from guilds to achieve game goals. These involve high-level players, making it difficult for a researcher to play alongside them without having put in many hours of game play. The time involved to become a legitimate member of a guild at this level may be impractical, and there is always a danger that the researcher as virtual ethnographer will cease to be self-reflective, becoming normalized to many of the cultural factors that they originally hoped to report on. Many researchers who are immersed in the *WoW* at this level have converted a private love of the game into an academic pursuit that can lead to biased work that expends too much effort on defending the area as a legitimate research concern rather than closely analyzing activity. The methodology I employed avoids these pitfalls as the researcher does not need to enter the game directly.

10.2.2 Retrospective Virtual Ethnography

This method involves the close cooperation of a guild member, as face-to-face interview and video screen capture are both crucial. Once a suitable subject is found, an informal interview is conducted to learn his position within the online community and his general relationship to the game, then interviewing the subject while observing him play. The advantage of being in the same RL room as your subject at this point is that it is possible to capture his commentary on game events and the actions of other players that would not be apparent remotely via the game. Discovering the methods a player utilizes to mediate between his role in the game and himself as an RL individual is a useful source of data.

At an appropriate point, a screen capturing system should be put in place. This may be as simple a pointing a video camera over the subject's shoulder, but it is much more effective to use a screen capture software installed on her computer as this will retain the legibility of the text chat channels. This method can also synchronize audio from the player and the game, so that the relative position of text and voice communication can be assessed, with "off-mic" comments made by the subject during play being captured.

It is ethically important for your subject to make it clear to other members of the guild at this point that you are capturing material in this manner. If the subject is willing, then he can capture the screen at what he considers significant moments during play when the researcher is not present. The risk with this method is that the subject will only capture what is important in gaming terms, such as raids, and may miss events that are valuable in research terms, such as informal communication between players.

The data can later be analyzed with a view to forming further interview questions that are framed to clarify actions and events in the screen capture videos. Subsequently reviewing the screen captures with the subject while moving through the questions that have come to light brings a useful perspective to gaming activity that may not have been apparent to the subject when she was fully immersed in game play. This perspective will often highlight the complex social interplay that is taking place during any collaborative activity and is likely to go beyond a simple functional description of the steps being undertaken to achieve a game goal.

10.2.3 Using Quantitative Data

Blizzard Entertainment allows players and researchers to gather a wide range of data from the game, including guild membership, time in-game, the density of players in a particular location, and the availability of certain objects. Used carefully, this qualitative information can provide an effective foundation to hypothesize about the nature of game play and communities within *WoW*. However, the complexity of the game is such that interpreting players' motivations and activity using only these data can result in misleading conclusions. Therefore, studies that draw on similar game data also tend to conduct a series of in-game interviews to balance the quantitative with the qualitative. A good example of the is "Alone Together" (Ducheneaut et al. 2006), which attempts to make the case that much of a player's time is spent alone while in-game. However, an assertion of this sort is difficult to support as colocation (being in the same part of the game world together at the same time) and a sense copresence are distinctly different. The chat channels and voice communication allows players who are "at a distance" in the world to socialize without having to account for the virtual geography of the environment.

Ultimately, an effective understanding of an MMO such as *WoW* can only be gained via a multidisciplinary approach and requires the researcher to take into account data and reasoning from outside his specialism. The thinking in this chapter is based on an analysis of a series of retrospective virtual ethnographic interviews conducted with guild members (White 2006) and supported by a range of emerging research literature on *WoW*.

10.3 The Lifecycle of a Player: The Linear View

A new player who knows nothing of the game and has no preexisting relationships with other players may be seen to go through three main stages of engagement. These stages can be described as infancy, childhood, and adulthood.

10.3.1 Infancy

During infancy, the player must come to understand new forms of communication, appropriate behavior and the rules of play. She learns through instruction, enquiry, and observation. Much of her time is spent exploring, experimenting, and improving her skills through trial and error. She completes simple quests especially designed to aid her understanding of the game world.

10.3.2 Childhood

In childhood the player comes to understand her role relative to others. Self-image and identity within the game world become more important. She starts to build social and instrumental relationships as small, shifting groups called "pick-up groups" (pugs) become required to tackle larger challenges. Communication develops from being necessary to gain specific problem-solving information and becomes increasingly social in nature. Tactical play-based decisions start to involve the assessment of the value of possible collaborations with other players. During this time her skills improve and his status increases as she earns trust from those around her.

10.3.3 Adulthood

In adulthood the player is likely to select a guild to join. This is a more formal group that exists to collectively achieve what cannot be completed by an individual player. The player assumes a role within the guild and understands the importance of reliability, punctuality, and skill. Through her membership, she works within a hierarchy to receive the highest rewards the game can bestow. She earns respect from her peers and gains kudos from other players. Some then go on to become advisors to the next generation, mentoring those who have taken on roles of responsibility.

Although this notion of a player's "life-stage" is useful because it gives an important sociocultural focus, mapping a player's movement through *WoW* in this manner implies an overarching, linear progression. The reality is far more complex, as the motivations and activities of players outlined within these stages tend to be in effect to a greater or lesser extent concurrently. For example, the guild leader will be constantly learning new skills and new forms of communication within a shifting social group. Similarly, many individuals play with friends they know well in RL and do not have to immediately build trust or form new relationships in the game. However, it is possible to segment the activity in and around *WoW* into a number of categories that, although they do not have sharp boundaries, are helpful in assessing the game from a social perspective. To this end, the following sections are loosely termed "Residency," "Relationships," and "Responsibility."

10.4 Becoming Resident: The Inculcation of New Players

What may start as a tentative foray into a computer game, for many becomes an immersive experience that leads to a feeling of belonging. This is the process when a player is becoming resident in the game world and is discovering that what he values about *WoW* has gone beyond a simple desire to complete game challenges.

10.4.1 Character and Role

Before entering the 3D world of *WoW*, the new player has to select a faction, a race, and a class of character. As she sifts through these options, a range of bizarre-looking creatures appears on screen.

At this point the player's choice may seem unimportant as he is likely to see the character or "avatar" she is choosing as simply a tool that he will control to complete game challenges. Each character comes complete with a short backstory and an introductory narrative that sets the scene and gives a basic motivation for the character to engage in the game world. The player is then invited to name this character and immediately discovers that all the most obvious names are taken, including versions of their own RL name. Depending on the player, the name chosen will usually be something that seems to fit the fantasy genre or something comic. A combination of these factors indicates that the avatar does not directly represent the player and invites the player to consider that she is playing a role that is set within the context of the overall narrative of the game.

As the player progresses through *WoW*, she may come to use her avatar as a chance to express aspects of her identity not normally visible in RL. She may choose to role-play (rp) an entirely separate fantasy character, or she may remain distinct from her avatar, simply using it as a function of the game. The spectrum of identification and rp is broad in *WoW* and catered to by the three types of game servers it is possible to logon to: rp, player-versus-player (pvp) and player-versus-environment (pve). Those that choose rp servers generally attempt to "stay in character" at all times whereas those on pvp servers might simply want to battle. The majority of players are somewhere in between these two extremes and will shift from playing the role that their avatar encourages to talking from an RL point of view about their day-to-day activities. Players become increasingly attached to their avatars as the game progresses. Others in the game start to assign kudos, respect, and trust to an avatar, and as such it begins to transcend the nature of a puppet, morphing into an embodiment of the player's commitment to the game and the visual home of their in-game relationships.

10.4.2 Leveling

The aim of the game in *WoW* is exceptionally simple: Complete a series of challenges to move through a series of "levels." Currently, following the recent *Wrath of the Lich King* expansion, the game now has eighty "experience levels" that are assigned to the player as he progresses. The design of the game is very elegant in this regard, requiring an increasing amount of playing time to be invested by the player as they move up the levels (Ducheneaut et al. 2006), with each level yielding a slightly greater prize, game currency, or game "gear" in the form of armor or weapons. The level a player has achieved is apparent within the game in a statistical form but is also made visible by the "gear" the character is wearing, which is indicative not only of the player's level but of the time and effort they have invested in the game (hence "experience" level.) In effect, gear is a visible form of social capital (Boride 1986). A high-level piece of gear, such as a powerful sword, embodies the skill of the player who owns it and the investment he has chosen to make in the game in terms of time. In addition, some high-level gear is tinted green or purple, giving a further indication of the type of play (adversarial or collaborative) that has been undertaken to gain the item. The way in which gear can represent skill, commitment, and play style is a significant factor in a player's decision as to who to "group" with. To a certain extent, it is possible to "judge a book by its cover" in *WoW*. The social capital of this gear is generally stable because the game is rigorously designed and constantly adjusted to try to ensure that there are no loopholes or shortcuts in leveling and the collection of gear. So, as a new player moves through the game, the gear they collect or buy becomes more than a mechanism to progress to the next level and starts to embody skill and effort. Often it embodies sheer commitment, as many quests and income-generating activities involve monotonous repetitive work known as "grinding." Players regularly complain about elements of the game design that involve grinding, but motivated by the clear rewards it brings they spend hours searching for rare objects or manufacturing revenue-earning products. Grinding is an intriguing example of where play can become work and is only engaged in because of the larger goals that are at stake.

Initially, leveling is a "side effect" of completing the early quests set by the game, simple tasks, such as collecting items or running errands. The early quests bring almost instant gratification (Rettberg 2008), rewarding the player with gear that will aid her path to the next level. A new player will also discover that the game world has areas populated by different nonplayer characters or "mobs/NPCs" at certain power levels, effectively making it impossible to explore the geography of the world without gaining the level required to fight these computer-controlled monsters. The simple process of moving through quests, and therefore levels, gradually becomes infused with an understanding of the social capital within the game. A player at level 4 will already have a sense of the effort required to reach level 20. As such, when a level 20 character strolls by she have will have an intrinsic respect for that player and the knowledge that she will have seen parts of the world unavailable to her level 4 avatar. The player becomes inculcated by the mechanics of the game as she comes to trust that each level requires a nonnegotiable amount of effort. So completing quests shifts from simply being a personal marker of progress to becoming an emblem of social capital that is on display for others to see. It is this aspect of the visual nature of the environment that has contributed to the success of 3D MMOs over and above their text-based counterparts.

10.4.3 Awareness

The multiplayer functionality of *WoW* is clearly what drives the social aspects of the game. However, understanding that you are playing on a server with thousands of other people and getting a sense of being copresent with other individuals has a significant impact on the players experience. If a new player does not have any preexisting relationships with other players, then she is likely to start to sense others via "ambient" awareness mechanisms within the game world. This ambient awareness is crucial to not feeling isolated and is one of the ways in which individuals feel a loose sense of belonging to a larger society of players. The effect can be likened to walking down a busy city street: The individual is unlikely to have any personal relationships with the people bustling by but he does feel connected to a larger social whole embodied by the city itself (Börner and Penumarthy 2003). The most obvious example of ambient awareness is the knowledge that the other avatars on screen are being manipulated by individuals at similar computers around the world. It is also comforting to realize that these fellow players are probably struggling with similar challenges, located in a geographical area in the game that is representative of a particular level or difficulty of play. This results in an innate companionship even if no direct communication is attempted.

Another significant form of ambient awareness comes via the "general" chat channel. This is a text chat channel that in essence broadcasts across a large area or zone and can therefore be seen by a large number of players. Even when being ignored by a player, the movement of the chat scrolling upward on screen is an ambient indicator that other people are interacting within the game world. This sense of social activity, however irrelevant to the player, gently reassures her that the world is genuinely populated by other human beings. There is an underlying understanding that the player may join in the flow of the general channel chat conversation at any point and that this could potentially be read by a large audience. The broadcasting style of this channel chat is similar to functionality in other social networking technologies such as Facebook™ (www.facebook.com) and Twitter™ (www.twitter.com). These ambient forms of social presence allow the new player to "lurk," drawing the new player into the social side of *WoW* without the need for immediate interpersonal engagement. Although the first communication with a fellow player may not take place until hours of gaming have been undertaken, a sense of playing with others will have been felt from the moment she first logged on.

10.4.4 Communication

There are many forms of communication in and around *WoW*, including the appearance of a character and the gear it has on, text chat, voice chat, the forums on *WoW*-related sites, and videos of play on YouTube. Of these, the predominant form is the in-game text chat. To a new player with little experience of the conventions of "instant messaging" on the Web, the form this text chat takes will be complex and mysterious, involving a myriad of shorthand phrases, some of which are generic to text speak and some of which refer specifically to *WoW* (Yee 2008). These phrases have developed in an attempt to make text chat more efficient and to allow emotional inferences to be made quickly (Moore et al. 2007). They have become essential communication tools in both negotiating the game and in maintaining social relationships [see Steinkuehler and Williams, 2006, for example, for the generic use of :)] to show that a comment is lighthearted. These include the abbreviation of place names and the more gaming specific cheer, \o/. Experienced players are likely to string together a series of abbreviations as they converse, in some cases constructing short sentences almost entirely of text speak that are incomprehensible to the uninitiated.

Fortunately, as previously mentioned, because it is possible to view the flow of chat in some of these channels without contributing, a new player can silently pick up much of the phraseology before taking the risk of making himself heard. In the later stages of the game, the ability to communicate efficiently in this manner can be the difference between success and failure when battling a powerful foe, which is why the density of abbreviations in a player's text-based dialogue is often an indicator of his experience. Despite much of the etiquette and phrasing being shared with text messaging and instant messaging, this form of communication contributes toward *WoW* feeling like a unique world, a culture apart from our own with its own language. For the new player, this can be alienating; but as this mode of communication is learned, it heightens the sense of belonging to an exclusive culture (Carr and Oliver 2009).

10.4.5 Ethos

In addition to fact-finding, tactics, and strategy, the new player also needs to gain an understanding of the games ethos (Nardi et al. 2007). Just because an action is permissible within the game's design does not mean that it is acceptable within the culture of the game as mediated by the players and may be considered to be deviant play (Mortensen 2008). Two examples of this are ganking and tweaking. Ganking occurs when a player waits for another player to kill a foe then swoops in and kills that weakened player near the end of the battle and takes over for an easy win. Twinking is when a high-level player equips a lower level player with gear she could not otherwise obtain. Neither of these activities is against the rules in the way that selling in-game currency for RL dollars is, and yet they are frowned upon by players as not being in the spirit of the game. In the same way that much of what constitutes socially acceptable behavior in RL is not codified as rules or law, the ethos of the game is not explicit. *WoW*, like many multiuser or multiplayer online environments, supports a culture that goes beyond the computer code the designers have put in place. This spectrum of "legitimate play" (Carr and Oliver 2009) is under constant negotiation and has, at one extreme, players pitting themselves solely against the game as designed and, at the other, rp in which individuals would prefer to lose a battle to stay in character.

Power gaming is a practice in which the player is only concerned to move up through the levels and gain the best gear as quickly as possible. These players are less likely to engage socially and can be seen as arrogant and disruptive (Taylor 2003). Depending on what a player considers legitimate play, power gaming may seem more like work than fun, as it involves hours of grinding and complex statistically based tactics. The more sociable side of the game can also become disruptive when issues such as not being invited into a guild give rise to a large amount of discussion that is not related to game play (Nardi and Harris 2006). In both cases, there is tension between the desire to complete game goals and the expectation that players should act like citizens within the world of *WoW*, taking time to integrate socially.

10.5 Forming Relationships: Integrating into a Society

As the new player progresses through the early levels and gains an understanding of the ethos and culture of the game, she may start to form relationships with other players, which can occur as she asks for help and advice or as she becomes skilled enough to help others. These relationships can be managed by using a basic type of social networking functionality embedded in the game, in the form of the friends list, that an individual can use to mark others she trusts as friends. The list will then indicate which of her friends is online at any one time, making it easy to make contact through the game even when no formal meeting time has been arranged.

10.5.1 Preexisting Relationships

So far we have made the assumption that the individual has entered the game on her own and has no preexisting relationships with other players. However, it is notable that roughly one third (Yee 2006) of players extend existing relationships into the game world, including RL couples, family groups (Nardi et al. 2007), and peer groups, such as college friends. In these cases, the notion of the "semipermeable membrane" (Castronova 2005) between the game world and RL becomes relevant, as ideas and conversations span the real and the virtual. The game becomes another location in which these relationships are played out, which can lead to new dynamics between individuals, such as the mother and daughter who play as equals (Steinkuehler and Williams 2006) or can simply maintain relationships between friends at great distances.

These preexisting relationships, especially family groups, are likely to have strong ties that involve emotional support and accountability. Putnam (2000) describes this as "bonding" social capital indicative of deep relationships that can be exclusive in nature. In contrast, the majority of relationships that are formed in WoW, as opposed to relationships that are continued in WoW, tend to have weak ties and are marked by "bridging" social capital. These relationships tend to be less serious and more inclusive and to allow a mix of individuals from diverse backgrounds to form loose groups. In this way, these less intense relationships are likely to broaden an individual's outlook and provide a potential pool of skills and knowledge for her to draw upon.

The class system in WoW is probably the most explicit form of social engineering in the game's design, and encourages the formation of informal relationships. Each player chooses to play a race (orcs, gnomes, trolls, humans, etc.), and within these races he chooses a class (Druid, Hunter, Rogue, etc.). Significantly, each class has a different range of skills, and as the game progresses a cross section of these skills is required to complete challenges. A successful group will need a combination of aggressive military-style classes such as the Warrior to fight mobs directly and healer classes such as the Priest to give characters support as they are attacked. Around level 16 the game introduces the next layer of quests in which a player will discover that she needs help from others to progress as he attempts a range of "dungeons" containing enemies that cannot be defeated by a single player. Some players will be content to form ad hoc groups with others as the need arises, whereas others will prefer to join more formal clubs of players to reduce the risks inherent in grouping with strangers.

10.5.2 Pick Up Groups

Dungeons, or "instances" as they are known, are private spaces that allow groups of between five and ten players to battle a series of mobs who will leave valuable "loot" when killed. The requirement to form small groups, coupled with the effectiveness of a team consisting of a range of classes, forces individuals to cooperate and to form new relationships. The pick up group or "pug", as it is known, is the informal mechanism by which players will gather to tackle dungeons. These pugs act in a similar manner to Engeström's "knotworks" (Engeström et al. 1999), small groups who form briefly to perform a specific task after which they disband. A pug is created when players stand in populated areas or near the starting point of an instance and use the text channels to either find a group or form a group made up of an effective range of classes. These groups often include strangers who know nothing of each other beyond the information the game provides directly, such as level and class.

The need to cooperate and work as a team to complete dungeons, combined with the relatively casual formation of pugs, leads to interesting social situations in which trust can be built and new relationships formed. However, tensions can arise between players who have different expectations in terms of friendliness, sharing, leadership, or roles (Williams et al. 2006). Similarly, players can quickly gain a reputation for not being a team player by communicating ineffectively or not showing sufficient focus at times when the pug needs to cooperate. The risk that a player will waste time and become frustrated if she becomes part of a weak pug is one of the reasons that guilds form. This will be discussed later in the chapter.

10.5.3 WoW as a "Third Space"

The designers of *WoW* understand that one of the main factors that encourage players to keep returning to the game is the chance to socialize informally and maintain relationships, designers tending to view MMOs as community platforms and not simply games (Ducheneaut et al. 2006). The casual bonhomie that *WoW* seems to facilitate has led to the successful description (Steinkuehler and Williams 2006) of the game as a virtual version of what Oldenburg describes as a "third place" (Oldenburg 1999). In simple terms, third places are neither work nor home but informal spaces on neutral ground that have an inclusive membership that values playfulness and wit; good examples of RL third spaces are the sports club or the pub. In these spaces, society is leveled and individuals are not ranked by their standing in the wider world. To a certain extent, this is true of the midlevel part of *WoW* in which players have a shared understanding of the game and mix relatively freely, forming pugs when necessary and logging in and out at their leisure.

It is worth noting at this point that players with a level difference greater than 4 cannot group. In effect, if an individual spends less time playing than his friends, he will fall behind, and the next time he logs on he will find he cannot be involved in the same activities. This has the potential to erode relationships, as opportunities for shared in-game experiences cease to be available, and acts as a major social driver for players to spend more time in the game and not get left behind, a lonely and ignominious position to be in. Whether this is seen as a deliberate ploy or an unavoidable effect of progression in a multiplayer environment, *WoW* is manipulating a social motivation to maintain relationships in an attempt to increase playing time. This is mitigated to a certain extent by a number of features that allow lower level players to catch up, thus softening the negative social effect of the games structure, helping to retain less committed players and broadening *WoW*'s appeal (Ducheneaut et al. 2006). Again, there is a fine balance between social and game-based motivation that has to be managed by Blizzard.

10.6 Dealing with Responsibility: Contributing to a Community

10.6.1 The Rationale for Guilds

The later stages of the game, starting at around level 40, involve an increasing amount of challenges that cannot be completed alone. These challenges come in the form of ever more complex dungeons or, at the apex of the game, raiding, which can involve groups of up to twenty-five players. To ensure that a sufficient number of trustworthy players of the right skill level are available at any given time to form a group, players must be drawn from a large pool of potential collaborators. The key phrase here is "trustworthy." The challenges at the end of the game are too sophisticated and too time consuming to risk tackling them with a casually formed collection of strangers. Players at the later stages of the game need to know of each other's strengths and weaknesses; they also need to use a more intuitive form of communication that can only develop over the course of hours of cooperative game play.

Groups of players who form to this end are called "guilds." Around 66% of *WoW* players are in a guild; this increases to 90% for players over level 43 (Ducheneaut et al. 2006). The size of guilds varies from fewer than ten to more than two hundred. However, guilds with more than 150 members can be tenuous and have many fringe members. This is in line with the Dunbar constant (Dunbar 1993), which suggests that the maximum membership of a stable community in which every member knows each other and trust is built is around 150.

Within the guilds, there is often a core group of around six players, with the larger guilds containing multiple core groups of a similar size. Most guilds will have some sort of "home" outside of the game, such as a Web site that displays guild members' statistics harvested from the *WoW* database and forums to discuss group activities. Again we see here the "semipermeable membrane" (Castronova 2005) that encases *WoW* as the game spills out onto the Web. *WoW* has an almost symbiotic relationship with the guilds, which although they are not mandated by the game are provided with certain specialist functionality, such as the guild chat channel.

The existence of guilds points again to the tension between game play and socializing. It is functional in the sense that it exists as a response to the challenges the game presents, but it is also social as individuals at this stage of the game need to know that they are in a team with likeminded players. With regard to this, guilds tend to present themselves has having a particular ideology or approach to the game. For example, a guild might claim that its primary goal is to ensure that members "have fun," possibly at the expense of being successful in game terms. Conversely, a guild that is tied to the power gaming ethos might advertise that it will take no "time wasters" and have stringent statistical requirements for membership. One particular guild requires all members to wear a pink tabard with a picture of a bunny rabbit skull on it (White 2006); this guild is almost as likely to visit the in-world pub as it is to attempt a challenging raid. However, this picture is not as polarized as it may appear on the surface, as members of even the most militaristic of guilds describe their reason for joining as being as much a social choice as a tactical one (Williams et al. 2006).

10.6.2 Modding

Another inherent tension that surrounds the game is the practice of "modding," an intriguing conjunction between the community and the game design elements of *WoW*. Blizzard keeps the access to certain aspects of the game's computer code open to players, which allows those with development skills to create modifications (mods) to the game itself, which are usually freely shared. Normally these are in the form of refinements to the user interface, for example, more detailed updates on other players' health status for raiding parties. Occasionally the player community creates a mod that damages the balance of the game, such as automatic targeting, at which point Blizzard releases an update that attempts to quash this new form of "illegal" behavior. Conversely, if the mod is generally perceived to be of use to the majority of players, then Blizzard will build it into the next official release of the game, thereby evolving the game platform in line with the community's wishes.

This process of development that involves the community is essential for the longevity of all online social spaces and is in evidence in different forms with major platforms such as Facebook™ and Twitter. In *WoW*, modding is a key example of the conversation between the game's designers and the players as they negotiate the boundaries of what constitutes legitimate play, the definition of which is in constant evolution as the game provides new challenges and the players invent new techniques to succeed using the code provided. This mode of engagement requires constant vigilance by Blizzard, forcing them to act as both encouraging stewards and authoritarian police over the communities within their game world.

10.6.3 Guilds as Communities?

Leaders and members within guilds generally have an understanding that there is a need for "a certain level of maturity, responsibility, and player welfare" (Williams et al. 2006). This awareness of the social needs of players goes beyond what is functionally required of the game; some guilds will go as far as to split the "pastoral" and the "functional" leadership roles between a guild leader and a raid leader—the guild leader acting as a form of matriarchal/patriarchal head of a community, managing disputes, maintaining morale, and overseeing membership, while the raid leader plays the role of a military general, giving orders during the heat of battle.

Given the nature and the makeup of guilds, it is possible to view them as a form of online community or "affinity group" (Gee 2003). More precisely they can be described as a "community of practice," a group of individuals with a joint enterprise, a shared repertoire, and mutual engagement (Wenger 1998). A successful guild will negotiate the game while maintaining relationships, striking an appropriate balance between leisure and commitment as laid out in its mission statement.

10.6.4 Commitment

It is the concept of commitment, not only to the game but to others within the guild, that starts to shift the nature of *WoW* away from that of a "third space." Complex raids require planning, including meetings, to discuss roles, responsibilities, and tactics before entering a designated battleground. Raids can take hours to complete and require that each member of the raiding party be in the right place at the right time with the correct equipment. In addition to this, raiding can be laborious. When I first suggested to a guild leader that I would like to watch him raid, he replied that he had a very specific role and would "mainly be pressing the F3 key for two hours" which was the method he used to heal the "tank" character that focuses the wrath of the main mob. Often a raid will be undertaken to supply a member of the guild with a specific powerful piece of gear that the mob drops when defeated. It is commonly agreed in advance that the loot from a raid is split disproportionately between the members of the raid party. This means that the motivation of some of the team is not to receive rewards proportionate to their effort but to work for the "greater good" of the guild.

As players enter into these later stages of the game and join guilds, their social responsibility toward fellow members can shift the "come and go as you please" nature of the game to one that can feel less like visiting the pub than going to work. The negative aspect of this is that guild members who meet in-game can assume some of the responsibilities that a relationship with strong ties have, without necessarily receiving the benefit of the emotional support that "bonding" social capital can bring. Up to 5% of players who form relationships in guilds bond to the extent that they choose to meet in RL; however, the majority who do not have preexisting relationships consider their relationships with guild members as casual or inconsequential (Williams et al. 2006). The social obligation members feel to spend time in-world can mount, especially for guild leaders, and a sense of duty rather than an opportunity to relax can become the reason for logging onto the game.

10.6.5 Guild "Churn"

Ultimately, *WoW* is more akin to a third space than to an office or a family home though leaving a guild or leaving the game obviously does not have the same implications as walking out of a job or getting a divorce. The relatively minor RL effects of quitting shifting social groups in-world accounts for the high "churn" rate in guilds, many of which disband within a month of formation. A successful guild requires strong leadership with a level of commitment similar to that of running an RL club or society. Many guild leaders become disillusioned with the constant tending of their guild and drop back into the ranks or leave altogether. A guild leader in one incident left a large guild to form a much smaller one made up of people he had long-standing relationships with. He described this process as "like leaving a busy student bar to go to your local pub for a chat with friends." This flux in leadership within a guild can also have positive effects, as it prevents hierarchies from becoming stagnant, maintaining a fluidity that is required to keep the community vibrant (White 2007b). For some, the chance to lead a guild is an opportunity they may not have in RL and is an exhilarating form of rp in itself.

10.6.6 The Informal Defined by the Formal

Attempting raids is an intense collaborative activity that is the core focus for most guilds. In *WoW*, the periods before and after a raid, or while waiting to form a pugs, are some of the most sociable moments. Before a raid, there will be a period when everyone gathers, during which a discussion of tactics and roles will take place; numerous crosscutting private conversations will also occur via the "whisper" chat function and the array of chat channels. Some of these will be simple greetings, enquiring after friends; some may be more divisive in nature, such as discussions over the suitability of a player for the raiding party or news of other guilds' exploits. Similarly, after the foe has been vanquished and the loot is being divided, an air of camaraderie, banter, and good-natured sarcasm will fill the chat channels, possibly ending in a trip to the in-world pub or a celebratory dance. The formal activities that the game provides give a center for the clustering of informal social activity. The community of the guild would not exist without the joint enterprise, a shared repertoire, and mutual engagement (Wenger 1998) facilitated by the game. In this way, *WoW* provides the correct ecology for the society of the guild to thrive in.

10.6.7 Alts

An intriguing and important aspect of *WoW* is that it allows players to create alternative or additional characters called "alts." This allows players to experience different races and classes, and it allows them to effectively restart the game in parallel to their main character. An interesting effect of allowing alts is that many players in the later stages of the game will start an alt to be able to enjoy the less complex, more "fun" areas of the game that have become tactically irrelevant to their main character. Therefore, playing an alt can be a relaxing break from the responsibilities of high-level play and a chance to help or mentor lower level players. Often an alt is created to stay in contact with a group of friends who play less regularly and cannot group with the level of the main character. The prevalence of alts demonstrates the multidimensional nature of *WoW*, in which the drive to progress through the levels to the endgame is only one of a myriad of possible motivations to log on.

10.7 The Importance of MMOs as Indicative of Society Moving Online

MMOs such as *WoW* are leading the way in the appropriation of the multicon-nectivity of the Internet for game playing and, more importantly, the facilitation of online communities, which is crucial to their long-term success. In addition to this, the MMO is a useful indicator of the inherent need in contemporary society to reconnect at an interpersonal level after decades of atomization. As Putnam (2000) demonstrates, we have been "bowling alone" for too long; and this is now fuelling an appetite for building social capital using online platforms.

The "macrofellowship" that the Internet is capable of supporting between large groups is heralding a new era, in which elements of individuals' engagement with society are increasingly situated online. This effect is redefining what it means to be a citizen, expanding the concept into an individual's physical *and* digital activities, with the potential for codes and norms to be constructed beyond traditional cultural and nation-state boundaries. Understanding the manner in which the society within an MMO functions gives an insight into some of the new paradigms the citizen of the near future will be engaging with.

Ultimately, despite its hundreds of millions of subscribers, an MMO such as *WoW* will always be a niche subculture both on and offline, because the genre it projects is extremely specific and not to everyone's taste. Nevertheless, the complex relationship between the game as a technological platform and the communities it supports is indicative of patterns of behavior, which then trickle down through other, more mainstream services on the Web, such as social networking platforms or microblogging. Platforms that are less immersive, and are therefore easier to mold around day-to-day commitments and responsibilities, are becoming more normalized within society than the dragon-slaying antics of the MMO. Nevertheless, it is helpful to have first studied the behavior of players in their complex, game-based, counterpart to be able to arrive at an understanding of how users act in these increasingly domesticated platforms.

10.8 Conclusion

In building an understanding of an MMO such as *WoW*, both the MMO as a game and as a community have to be taken into account. This requires an understanding of the rules and mechanisms put in place by the designers and how these both facilitate and are influenced by a culture made up of millions of players. *WoW*'s success is in part due to its initial design as the latest incarnation of a form of gaming with a heritage that spans decades. More importantly, Blizzard Entertainment understands that as in any culture, the boundaries in *WoW* are under constant negotiation. Players will push these boundaries, on the one hand, and demand an equitable playing field, on the other. Blizzard is constantly assessing this using statistical sources and information from community liaison staff, allowing the company to make regular subtle adjustments to the dynamics of the game in an attempt to ensure that the balance between reward and challenge and between cooperation and competition is maintained.

WoW can be enjoyed at a myriad of levels: the satisfaction in progressing through the game, the pleasure of socializing with friends and meeting new people, the sheer audiovisual spectacle, the sense of being immersed in an alternative world, the chance to develop new skills and gain the respect of fellow guild members, and the opportunity to rp and to leave RL restrictions behind. This range of motivations means that there is no single path to becoming a good player; it is possible to be respected as much for social or pastoral skills as it is for being at level 80 and showboating the most powerful gear. It is the breadth of possible activity and the understanding that legitimate participation in *WoW* can be much more than leveling that is instrumental in providing an environment that supports thousands of communities and subcultures (White 2007a).

In social terms, the design of *WoW*, the 3D visuals, the game challenges, and its complex communication channels all work together to draw a player in and give her a sense of being copresent with other individuals. Whether slaying a sixty-foot dragon in a team of twenty-five, fishing with friends while chatting about work, or dancing the conga through a marketplace, it is the knowledge that there is an audience of friends and strangers that lends value or social capital to each activity. The strength of *WoW* is its ability to facilitate this sense of presence while encouraging a range of both formal and informal activities. It does this within an overarching game structure that provides a shared focus and an opportunity for common understanding among a growing society of players.

References

Bourdieu, P. 1986. The forms of capital. In *Handbook of Theory and Research for the Sociology of Education*, ed. J. Richardson, 241–258. New York: Greenwood.

Börner, K., and Penumarthy, S. 2003. Social diffusion patterns in three-dimensional virtual worlds. *Information Visualisation* 2: 182–189.

Carr, D., Oliver, M. 2009. Tanks, Chauffeurs and Backseat Drivers: Competence in MMORPGs In *Eludamos. Journal for Computer Game Culture* 3 (1): 43–53.

Castronova, E. 2005. *Synthetic Worlds*. London, UK: The University of Chicago Press.

Ducheneaut, N., Yee, N., Nickell, E., and Moore, R. J. 2006. "Alone Together? Exploring the Social Dynamics of Massively Multiplayer Game. "*In Conference Proceedings on Human Factors in Computing Systems (CHI 2006)*, 407–416.

Dunbar, R. I. M. 1993. Coevolution of neocortical size, group size and language in humans. *Behavioral and Brain Sciences* 16(4): 681–735.

Engeström, Y., Engeström, R., and Vähäaho, T. 1999. When the enter doesn't hold: The importance of knotworking. In *Activity Theory and Social Practice* eds. S. Haiklin, M. Hedegaard, and U. Jensen. Denmark: Aarhus Press.

Gee, J. 2003. *What Video Games Have to Teach Us about Learning and Literacy.* New York: Palgrave Macmillan.

Moore, R., N. Ducheneaut, and Nickell, E. 2007. Doing virtually nothing: Awareness and accountability in massively multiplayer onling world. *Computer Supported Cooperative Work (CSCW)* 16 (3): 265–305.

Mortensen, T. E. 2008. Humans playing World of Warcraft: Or deviant strategies? In *Digital Culture, Play and Identity*, ed. J. W. Rettberg, 203–225. Cambridge, MA: Massachusetts Institute of Technology Press.

Nard, B. A., and Harris, J. 2006. *Strangers and Friends: Collaborative Play in World of Warcraft.* Irvine: University of California.

Nardi, B., Ly, S., and Harris., J. 2007. Learning conversations in world of warcraft hicss, 40th *Annual Hawaii International Conference on System Science (HICSS'07)*, 79.

Oldenburg, R. 1999. *The Great Good Place: Cafes, Coffee Shops, Community Centers, Beauty Parlors, General Stores, Bars, Hangouts, and How They Get You Through The Day.* New York: Marlowe and Company.

Putnam, R. D. 2000. *Bowling Alone: The Collapse and Revival of American Community.* New York: Simon and Schuster.

Rettberg, J. W. 2008. Quests in World of Warcraft: Deferral and repetition. In *Digital Culture, Play and Identity*, ed. J. W. Rettberg, 167–185. Cambridge, MA: Massachusetts Institute of Technology Press.

Steinkuehler, C., and Williams, D. 2006. Where everybody knows your (screen) name: Online games as "third places." *Journal of Computer-Mediated Communication 11*(4): article 1.

Taylor, T. L. 2003. *Play between Worlds: Exploring Online Game Culture.* Cambridge, MA: Massachusetts Institute of Technology Press.

Wenger, E. 1998. *Communities of Practice: Learning, Meaning and Identity.* New York: Cambridge University Press.

White, D. 2007a. Cultural capital and community development in the pursuit of dragon slaying. Paper presented at the Games Learning and Society Conference, Madison, WI. July 12–13, 2007. http://tallblog.conted.ox.ac.uk/index.php/2007/07/30/cultural-capital-and-community-development-in-the-pursuit-of-dragon-slaying/. Accessed November 28, 2008.

White, D. 2007b. Dave's top 10 musings on the encouragement of community in multi-user virtual environments. http://tallblog.conted.ox.ac.uk/index.php/2007/05/14/daves-top-10-musings-on-the-encouragement-of-community-in-multi-user-virtual-environments/ Accessed November 28, 2008

Williams, D., Ducheneaut, R., Xiong, L., Yee, N., and Nickell, E. 2006. From tree house to barracks: the social life of guilds in World of Warcraft. *Games and Culture* (4): 338–361.

Woodcock, B. Data on MMO subscribers. http://www.mmogchart.com/ Accessed on November 29, 2008.

Yee, N. 2006. The Demographics, motivations and derived experiences of users of massively-multiuser online graphical environments. *PRESENCE: Teleoperators and Virtual Environments* 15: 309–329.

Yee, N. MMO lexicon, http://www.nickyee.com/daedalus/archives/001313.php Accessed November 29, 2008.

Chapter 11

Perceiving an Internet Community as a "Utopia"
Beliefs, Norms, and Resistance among Older Chinese

Bo Xie

Contents

11.1 Introduction

During the past decade, there has been tremendous growth in Internet adoption in China. In October 1997, there were only 620,000 Internet users in China (China Internet Network Information Center 1997). Yet, by the end of

243

June 2008, 253 million Chinese people had already gone online (China Internet Network Information Center 2008). The dramatic growth of the Internet in China is coincident with the aging of the Chinese population. Data from the last Chinese census indicate that in November 2000, 6.96% of the Chinese population (88.11 million) was aged sixty-five or older (National Bureau of Statistics of the People's Republic of China 2001). In 2007, approximately 7.9% of the Chinese population (104 million) was aged sixty-five or older (Central Intelligence Agency 2007). It is projected that in 2030, 16.57%, or 243 million Chinese, will be aged sixty-five or older (National Bureau of Statistics of the People's Republic of China 2001).

At the intersection of the Internet trend and the aging trend in China is the constantly growing subpopulation of older Chinese Internet users. Although the percentage of older Chinese Internet users (age fifty and above) has been consistently lower than 5% of the total Chinese Internet population, because the sheer size of the total Chinese Internet population has increased so much, the sheer size of the older Chinese Internet user subpopulation has also increased significantly. In June 1998, there were only 14,400 older Chinese Internet users (China Internet Network Information Center 1997); yet, by the end of June 2008, almost 10 million older Chinese were already surfing the Internet (China Internet Network Information Center 2008).

Both the Internet trend and the aging trend in China have far-reaching implications for Chinese society and Chinese people's everyday lives. Yet, to date very little attention has been paid to the intersection of these two trends. As one of the first attempts to look into this intersection, I conducted a study to examine the impact of the Internet on older Chinese persons' civic engagement, social relationships, and psychological well-being, and compared the findings with those from an examination of the Internet's impact on older American Internet users (Xie 2006a). Although this study covered a broad range of aspects and many key findings have been reported elsewhere (Xie 2005, 2006b, 2007a, 2007b, 2007c, 2008a, 2008b, 2008c; Xie and Jaeger 2008), in this chapter I focus specifically on one key aspect of this study that has not been reported previously: how participation in a senior-oriented Internet community affects many older Chinese persons' beliefs and norms in contemporary China. Below I first lay out the Chinese historical, political, and economic contexts that affect Chinese people's aging experience. I then report the research site, participants, and research methods of this study. Next, I discuss one of the key findings of the study: The Internet community is a "Utopia" that helps older Chinese to preserve their beliefs and norms. Finally, I look into the ironies behind this utopian perception and explore the implications for Chinese society and the lives of older Chinese.

11.2 Aging in the Chinese Context

Compared with the aging populations in industrialized nations, the aging Chinese population has several unique characteristics: (1) the absolute number of older Chinese people is very large; (2) the growth of the older Chinese population is rapid; (3) unlike most aging societies, China has an underdeveloped economy; (4) population aging in China shows tremendous regional differences: The proportion of the older Chinese population in the more developed coastal regions is greater than that in the inland regions; and (5) the growth of the oldest-old subpopulation (people aged eighty and above) is notable, with a growth rate of 5.4% per year. This subpopulation had increased from 8 million in 1990 to 11 million in 2000 and is projected to reach 27.8 million in 2020 (Lee 2004).

A widely circulated view argues that in traditional countries such as China, seniors receive more respect and have higher social status than their age peers in industrialized societies (Palmore and Maeda 1985; Palmore and Manton 1974; Cowgill and Holmes 1972; Simmons 1945). For instance, Streib (1987) argues that American society, with its emphasis on self-reliance, independence, free choice, and self-determination, is advantageous for active older adults who are in good health and especially good financial situations, whereas Chinese society, mainly due to its norms of reciprocity and orientation of the family, is advantageous for frail older adults.

This view has increasingly been challenged because, during the past several decades, the forces of industrialization, capitalization, and globalization have resulted in dramatic social changes in China (Perry and Selden 2003; Price and Fang 2002; Bian 2002; Whyte et al. 1977; Walder 1989; Ikels 1996; Lee 1998; Tang and Parish 2000; Warner 2001; Zhou and Hou 1999; Zhou et al. 1997). As a result, the distinctions between Chinese and industrialized societies may be blurring. There is empirical evidence that in contemporary Chinese society, filial piety may be changing, perhaps even eroding (Ng et al. 2002; Liu 1994; Joseph and Phillips 1999). Due to the continued demographic transition and dramatic social and economic transformations (e.g., the decline in both fertility and mortality rates, the one-child policy, and economic reform), the traditional family support system in China is experiencing great challenges in maintaining its capacity to provide support for older adults (Sun 2002; Tu et al. 1989).

Research suggests that older Chinese persons' living conditions, health, and financial situations largely lag behind those of their age peers in the West (Chappell 2003; Li and Tracy 1999). Public systems of direct relevance to the older Chinese population, including the public health and welfare systems, have gone through dramatic changes in recent years. Yet, they are still underdeveloped and insufficient in meeting the challenges of the aging Chinese population (Chau and Yu 2001; Lee 2004; Lee 2001). Recent reforms in the economic arena also have had a significant impact on the lives of older Chinese people. In particular, the conversion of state-owned enterprises (SOEs) into shareholding corporations, coupled with the Chinese SOE managers' failure to adapt to market-oriented practices, has caused great financial losses to the newly converted corporations (Freund 2001). This has often inevitably resulted in the laying off or forced early retirement of workers of all ages but especially older workers. Price and Fang's (2002) examination of the interplay between economic reforms, especially the downsizing of SOEs and globalization (e.g., China's entrance into the World Trade Organization), and individual characteristics found that age and education were the most salient variables in determining the impact of economic reforms. Older workers with less education can be categorized as "the discouraged old," who are most vulnerable to the negative consequences of economic reforms (Price and Fang 2002).

When it comes to education, a significant number of older Chinese people are at a disadvantaged position compared with their younger counterparts, largely due to the negative impact of the Cultural Revolution (1966–1976). During that decade, millions of Chinese youth were unable to pursue their education because school systems at all levels all over China were literally nonfunctioning. Further, between 1967 and 1978, millions of Chinese youth were "sent down" to rural areas. This experience has had lasting and powerful influences on the subsequent life courses and behavioral patterns of the send-down individuals, including significant delays in marriage and childbearing, less advantageous choices in the urban labor force when they returned to the cities, and disadvantaged economic well-being (Zhou and Hou 1999; Jiang and Ashley 2000).

Chen and Cheng (1999) argued that both the send-down policy that has had dramatic effects on the life course of the older Chinese generation and the subsequent economic reform policy that has been in place since the late 1970s have likely required individuals to adapt to completely different value systems, behavioral norms, and social skills. There is empirical evidence that supports this argument. For instance, Lu and Alon (2004) reported findings of a large government survey project conducted in September 2000 in Shanghai. The participants of that survey study—the "white-collars"—featured what Lu and Alon (2004) termed "three high and one low": high education (four years college education and above), high income (monthly income around Ranminbi [RMB] 3,000), high position, and low age (twenty to forty). The findings of that study indicated that, as compared with the older Chinese generation that grew up in the planned economic system who valued "security, the prospect of the work unit (which means stability) and social requirement" and "selfless contribution to society" (p. 78), young, educated, "white-collar" Chinese people who grew up after the economic reforms that started in 1978 instead valued self-realization or personal development and personal economic rewards.

11.3 The Present Study

During 2004–2005, I conducted an ethnographic study of older Chinese Internet users who were members of a senior-oriented computer training organization headquartered in Shanghai, China (Xie 2006a). Shanghai is the financial and economic center of and the largest city in China. By the end of 2004, it had a population of 13.5 million. Among those residents, 2.6 million (19.3%) were age sixty or older, and 2 million (14.9%) were age sixty-five or older (Shanghai Research Center on Aging 2005). The senior-oriented computer training organization that I studied is called *Lao*

Xiao Hai, which is a widely used Chinese phrase that refers to active seniors (who are as energetic, enthusiastic, and curious as kids) and can be literally translated as "Old Kids." In addition to providing face-to-face computer training, OldKids also maintains a Web-based, free, senior-oriented online community in which older Chinese people can interact with peers via interactive services such as online forums, text and voice chat, and instant messaging (Xie 2008b). A key feature of the OldKids community is that the majority of its members interact with peers both online and offline in the "integrated OldKids Internet community" (Xie 2005, 2006a).

I conducted semistructured, open-ended interviews and participant observation at the OldKids headquarters, computer classes, computer interest groups (which members called *Dian Nao Sha Long*, or computer salons), and member-organized social gatherings. I interviewed thirty-three older Chinese people who were OldKids members, five OldKids executives and employees, five local government officials who were in charge of various levels of the Seniors Committees in Shanghai, and administrators of two senior-oriented nongovernmental organizations (NGOs). The total number of Chinese interviewees was forty-five (Table 11.1).

The thirty-three older Chinese participants ranged from fifty to seventy-nine years in age (mean age = 62.5).* Nineteen (57.6%) of these older Chinese participants were female, and fourteen (42.4%) were male. Twenty (60.6%) of these participants were college educated, five (15.2%) high school educated, four (12.1%) technical secondary school educated, and four (12.1%) middle school educated (Table 11.2). Considering that only 12.6% of Shanghai residents had four or more years of college education (Shanghai Municipal Population and Family Planning Commission 2003), this sample of older Chinese participants notably had a higher level of education.

All thirty-three older Chinese participants were retired, even though some of them were still in their early fifties.† They all had relatively good pensions. Their average monthly pension was about 1,500 RMB (approximately U.S. $183 at the time), which was almost twice as much as the minimum living standard set by the Shanghai government and almost 50% higher than the average monthly income of older Chinese

---

* Anthropologists have long recognized that the aging experience varies greatly across cultures (Fry 1999). Informed by anthropologists' observations, I deliberately chose to not recruit older Chinese or American participants based on any predetermined chronological age. Rather, I recruited older Chinese and American participants for the larger study (Xie 2006a) based on their self-identification of being an "old" person. As it turned out, the chronological age of the self-identified older Chinese participants were notably younger than of the older American participants (and what Americans would typically consider to be "old"). This finding in itself is interesting because it is a reflection of how old age is defined and understood differently in these two cultures.

† These participants were all retired despite their younger chronological ages (than what Americans would typically consider to be retirement age). This reflects the influence of recent social and economic changes on the lives of older Chinese — in particular, the reforms in SOEs and subsequent laying-offs and forced early retirement, as discussed above (Price and Fang 2002).

Table 11.1 Age, Gender, and Role/Position of All Chinese Interviewees

		OldKids Members		OldKids Executives/ Employees		Government Officials		Administrators of Senior-Oriented NGOs		
		Female	Male	Female	Male	Female	Male	Female	Male	Total
Age	20–29			1	3	1				5
	50–59	11	3			2	2	1		19
	60–69	7	4		1				1	13
	70–79	1	7							8
	Total	19	14	1	4	3	2	1	1	
Total		33		5		5		2		45

people in the urban areas of Shanghai (Shanghai Research Center on Aging 2005). It is important to keep in mind that this sample of older Chinese people was small and not a random one in that these older Chinese people were self-selected to participate in the OldKids organization (as well as this study), had more formal education, and were in better financial situations than the majority of their age peers in China. As a result, this population is not representative of the older Chinese population in general.

Participants were recruited using the snowball sampling technique. Most interviews were conducted at the OldKids computer classrooms where the computer class and salon activities take place. A few were conducted at the participants' offices, private homes, or other locations of their choice (e.g., a nearby park). In several cases where the participants could not meet face-to-face, interviews were conducted via telephone, e-mail, or instant messaging (for a methodological discussion

Table 11.2 Educational Background of the Thirty-Three Older Chinese Participants

		College		High School		Technical Secondary School		Middle School		Total
Age		Female	Male	Female	Male	Female	Male	Female	Male	
	50–59	5	2	2			1	3	1	14
	60–69	5	2		1	2	1			11
	70–79		6	1	1					8
Total		10	10	3	2	3	1	3	1	33

on the interview techniques used in this study, see Kazmer and Xie 2008). Each interview lasted about an hour; face-to-face and telephone interviews were recorded using a digital voice recorder. An informed consent form (in Chinese, approved by the Institutional Review Board of Rensselaer Polytechnic Institute) was completed by each participant before each interview was conducted. Major interview questions most relevant to the scope of this chapter included: What is your view of the Internet? And OldKids? Has using the Internet affected your life in any way? Has being a member of OldKids affected your life in any way?

Data analysis for this study was guided by grounded theory (Glaser and Strauss 1967; Strauss and Corbin 1998), such that data collection and analysis occurred simultaneously to ensure that emerging theory was firmly grounded in data while data analysis was strictly guided by theory. Detailed descriptions of the data analysis for this study can be found in Xie (2008a, 2008b).

11.4 The OldKids Internet Community as a "Utopia"

A novel finding of this study is that many older Chinese Internet users who participated in this study view the OldKids Internet community as a *Shi Wai Tao Yuan* or Utopia that is free of the "pollution" of market systems.* As the following quotations illustrate:

> Compared with the mundane world where everybody is insanely pursuing money and material pleasure, I think this [the OldKids community] is like a *Shi Wai Tao Yuan*. It seems that people here are less utilitarian. People like the three cofounders of OldKids, instructor Wu

---

* The phrase "Shi Wai Tao Yuan," as used in Chinese society, is roughly equivalent to the concept of "Utopia" as used in Western countries. This phrase consists of four Chinese characters, and the first character, "Shi," literally means *the mundane world*; the second character, "Wai," means *beyond* or *outside*; the third character, "Tao," means *the peach*; and the last character, "Yuan," means *the source*. A widely used Chinese—English dictionary interprets this phrase as "the Land of Peach Blossoms — a fictitious land of peace, away from the turmoil of the world; a haven of peace" (The Chinese-English Dictionary Editorial Team in the English Department at Beijing Foreign Language College, 1989. A *Chinese—English Dictionary*, p. 624. Beijing, China: Shangwu Publishing). This interpretation covers the basic meanings of Shi Wai Tao Yuan; to have a better understanding of this phrase, however, it is necessary to know its origin in the Chinese culture. The phrase Shi Wai Tao Yuan is originated from a widely circulated poem written by a famous Chinese poet — Tao Yuan Ming — about 1,600 years ago. Tao lived in the East Jin Dynasty, a period in Chinese history that was characterized by war, conflict, instability, and corruption among government officials. Tao himself was a low-ranking government official for a short period of time. He soon decided to quit his position because he felt that it was not worth to having to interact with those corrupt officials and with the corrupt political system. From that point on, Tao lived in a rural area, where he enjoyed a monetarily poor but free life. Tao wrote many famous poems about his poor but enjoyable lifestyle. One of these poems was about Shi Wai Tao Yuan, in which he described a fictional, ideal world — the Tao Yuan — where everyone lived in peace and harmony and did not have to worry about all of the problems that those who lived in the mundane world had to face.

Xiaofan, and the volunteers, they do not ask for money. They do not do things only for money. So I feel this is like a *Shi Wai Tao Yuan*. To those of us who have lived through that time, this is quite consoling. We finally have a Pure Land. People here love to learn new things. They are well educated and have the spirit of volunteering and dedication. So I feel this is very inspiring. It eases our soul. [Ms. H]

Learning computers at OldKids and being a member of OldKids have significantly enriched my life. I am especially impressed by the three cofounders of OldKids, who have selflessly devoted themselves to this community. I am very happy when participating in OldKids activities. I used to have many negative opinions about Web sites because, for example, some young people are addicted to playing computer games, and some students would even miss their classes to go to an Internet Café to play computer games. Things like these have made me think that young people should not go to Internet Cafés. But our OldKids Web site is different. I feel that it is like an unpolluted place, like a *Shi Wai Tao Yuan*. I think that the marketing behaviors [in the mundane world] are too heavily money driven. But here at OldKids, it's all about dedication, about learning. I feel that I have learned a lot from here. [Ms. T]

Similarly, another member describes the OldKids online world as a world that has "true feelings" because, unlike the mundane world where everything is about money, people in the online world are selflessly dedicated to helping others without asking for money in return:

I think the OldKids online world is wonderful. Although the online world is virtual, there are true feelings in this world. In the mundane world, everything is about money. The real-life world is a monetary world. You can't get anything done without money. However, in the online world, even though we may not know each other [in real-life], we all are selfless. Those who know more about computers teach others for free. I think this spirit is very valuable. For example, the first time when I visited OldKids' online chat room, I had no idea how to use it. Someone from another city, whom I never met before, tried very hard to teach me step by step. I was really confused by those steps, but s/he was very patient. S/he did not know me, and I did not know her/him, either. But they always welcome everyone who's new [to the OldKids online community] with open arms. It makes me feel that there are true feelings in this world. [Ms. Y]

Further exploration suggests that these older Chinese persons' perception of the OldKids community as a Utopia may be affected by factors associated with both the behaviors and norms of individual members and those of the OldKids organization. On the one hand, in the OldKids community, members willingly, selflessly, and enthusiastically help each other to learn—without asking for money in return, which is in sharp contrast to the market-driven practices in contemporary China. This peer teaching/learning process has also provided members with unique, rich opportunities to form meaningful social relationships that provide much needed companionship and emotional support for these older Chinese people. On the other hand, the OldKids organization has also played a—most likely unplanned—role in the formation of this utopian perception. It does so through its active self-promotion of being a "nonprofit" organization whose goal is to provide "community service" for older adults in Shanghai. Also, the three cofounders of OldKids are widely perceived by the majority of OldKids members as having a "spirit of dedication" or "spirit of selflessness," which certainly reinforces the perception that this community is not money driven.

Ironically, as will be discussed below, the organization's relabeling of being nonprofit is merely a strategic choice of seeking money to survive, and the three cofounders are straightforward in admitting their money-driven motivations behind all of their operations and practices. Why, then, do so many OldKids members still perceive the OldKids community as a Utopia? These issues are examined in detail below, through an overview of the history and development of the OldKids organization and key characteristics of the current OldKids community.

11.4.1 Learning and Relationship Building in the Oldkids On-/Offline Community

Peer teaching/learning and relationship building are accomplished both online in the OldKids online community and offline in the OldKids computer classes and salons. First launched in early 2000 (which was the peak time of the Internet boom in China), OldKids was originally a private, for-profit company targeting the older Chinese consumer market. Three young Chinese men who had just graduated from college started the business with their own savings and money borrowed from relatives and friends. Motivated by the worldwide.com fever, especially the success of Amazon.com, these three cofounders initially wanted to build an Amazon-like B2C (business-to-consumer) Web site targeting older Chinese consumers. After only several months of operation, however, they quickly realized that this business idea was not going to succeed anytime soon, simply because at the time there were not very many older Chinese people who knew how to use the Internet, not to mention how to use it to purchase goods. The dramatic bursting of the Internet bubble during that time also meant that the company had little chance of getting venture capital and thus had to survive on its own.

The cofounders decided to move from online to offline: that is, to focus on computer/Internet training for older adults (in the physical world) to cultivate the potential customer population of their Amazon-like Web site and, more immediately, to generate revenue from the training classes. By the time of this study, OldKids had trained more than one thousand older Chinese people in Shanghai to use computers and the Internet. OldKids computer classes typically take place at an Internet café where OldKids has rented two rooms (which are well equipped with the newest computer hardware and software, high-speed Internet access, and central air conditioning) for members to use from Monday through Saturday mornings.* Each OldKids computer class typically lasts four to eight weeks. During this period, students meet once per week to learn from the instructor, who is usually also an older person. To attract more students, OldKids has been keeping the class fee to approximately $15 to $30 per person, which is significantly less expensive than similar computer classes offered by other organizations.

In addition to the computer classes, OldKids also encourages and helps students of their computer classes to organize computer salons so that they can continue meeting with and learning from their peers after the completion of classes.† Although this practice does not directly generate revenue for the organization (membership is free), it is beneficial to the organization in that it helps to keep the students with the organization beyond the relatively short period of the class duration. And the students like this opportunity as well, because many of them feel that, after completing a computer class from OldKids, there is still a great need for continuous learning and practice (Xie 2007a). In fact, participating in an OldKids computer salon after completing an OldKids computer class has now become "a tradition" in the OldKids community.

There are several key differences between the OldKids computer classes and the computer salons. First, the former lasts for weeks, whereas the latter may last for years; second, the former provides opportunities for older Chinese people who have shared interests—but did not know each other before—to get together and start interacting and forming relationships, whereas the latter ensures the continuity of this interaction and relationship development. Third, the former features instructor-based training/learning, whereas the latter features peer training/learning (Xie 2007a). Engagement in such a peer-teaching/learning practice requires members' willingness and dedication in helping each other and sharing their computer knowledge and skills. And there is abundant willingness and dedication in this community, as the following quotations illustrate:

---

* The primary target customer groups of this Internet café are younger people, especially teenagers and those in their twenties. Because these customers typically do not go to the Internet café until at least after midday, this Internet café usually does not get much business in the mornings. Thus, OldKids was able to make a deal with the Internet café owners to give OldKids members a discount price (RMB 2 or approximately U.S. $0.25 per hour, which is only one third of the regular price) if they come and use the facility in the mornings.

† If students choose to meet at the same Internet café, they can get the same discount from the Internet café. This has been another good incentive for them to organize and participate in the OldKids computer salons.

> I participate in OldKids salon activities. The friends there are of great help to me. If I have any questions, I'll ask when we meet at the weekly salon meetings. We discuss each other's questions. If anyone knows the answer to my question, s/he will help me; and vice versa. The salon meetings are very good in helping us learn. [Ms. C]

> It's very common in our [OldKids computer] salon that, if you have any problem, someone will immediately help you solve it. We all feel that, once we've learned something, we just couldn't wait to teach it to other members of our salon. [Ms. S]

While the main focus of the OldKids organization has shifted to offline computer training, the OldKids Web site still functions. Considering the short history of OldKids and especially the low Internet use rate among older Chinese people, the OldKids online community is quite successful in terms of providing older Chinese people—especially current and former students of OldKids computer classes who live in Shanghai—an online environment in which they can interact with peers.* It does so by providing interactive services such as online forums, text, and voice chat, and instant messaging. My interviews with these older Chinese people and participant observation in the OldKids online community have revealed that each of these interactive services has primarily fostered one unique type of social support: online forums for informational support, instant messaging for emotional support, and online chat for companionship (Xie 2008b). The informational support—information and knowledge about computer skills, tips, and news—exchanged in the OldKids online forums is especially appreciated by these older Chinese people, particularly because they often have difficulties getting help with their computer problems from other sources, including their own adult children and other younger people (Xie 2007a).

For instance, one participant commented that, in today's market economy, he would not dare to ask a young person for help with computer-related problems if it might take more than ten minutes of the young person's time. This is because, he explained:

> In the planned economy era, I could have asked young people for any questions I might have about computers and they would have always been willing to help me, no matter how many questions I might have asked and how much time it might have taken to help me solve my problems ... But now [in the market economy era] more and more people care about themselves, while fewer and fewer people care about others ... [Mr. N]

---

* According to the vice president of OldKids, Wang Yong, by the end of 2004 OldKids had over 8000 registered online members in the nation, and about two-thirds of them were from the Shanghai metropolitan area. The geographic proximity ensures and greatly facilitates OldKids online members' face-to-face or offline interactions, thus promoting an integrated on/offline community.

In comparison, in the OldKids Internet community, members can always get help from others to solve any computer problems they may have—and for free. Many participants tell the stories of how members of the OldKids online community are always willing and eager to help others without asking for money in return, as the following quotations illustrate:

> I don't really have any particular person to contact [to seek help]; but if I have any questions about computers or the Internet, I can just post my questions in the [OldKids online] forum and many people will come and help me. I have seen too many people who did not know such things as uploading music; but now they all know how to do it. They all learned from others. [Ms. T]
>
> I did not know how to make birthday cards, so they [other members of the OldKids online community] started teaching me. I started from the basics. Many people in this online community have taught me. In this community, as long as you ask questions, somebody will answer you. In such an environment you will feel that, because everything I know is from others, if I know something that can be helpful to somebody, then I'd love to share with her/him, too. This is the most important thing I've learned in this community. I think that there are true feelings in the online community. The online community is not about money; it does not want your money. [Ms. Y]

In addition to the valuable informational support, members enjoy and appreciate the emotional support and companionship exchanged in the OldKids online community. While informational support is exchanged primarily in the OldKids online forums, companionship or interaction that is sought for purely social, enjoyable purposes (Rook 1995, 1987) occurs primarily in the online chat rooms. Emotional support is sought and received primarily via instant messaging, the private mode of computer-mediated communication (CMC) available in the OldKids online community. Details about these OldKids members' use and perceptions of these different modes of CMC have been reported elsewhere (Xie 2008b). The point of the most direct relevance to our discussion here is that the social interactions and relationships developed in the OldKids online community, and the informational and emotional support and companionship exchanged in this online community, may have all helped to shape these members' perceptions of the OldKids community as a "Utopia."

11.4.2 OldKids as a "Nonprofit" Organization with Three "Selfless" Cofounders

In addition to the learning and relationship-developing opportunities available in the OldKids online and offline community, another important factor that may have contributed to the participants' utopian perception is the OldKids organization's strategic choice and active self-promotion of being a "nonprofit" organization. This choice was made because although the offline computer training generates some revenue for the organization, it is still far from sufficient to keep the company running. To survive, OldKids has been actively promoting itself as a "nonprofit organization that provides community service for seniors" (Wu Hanzhang, chairman and cofounder of OldKids). This new position makes it much easier for the organization to gain support from the government as well as individual members. For one thing, interviews with several governmental officials at various levels and branches of the Shanghai government—especially those in charge of senior-related matters—reveal that the officials have widely acknowledged and positively responded to OldKids' nonprofit position.

The Shanghai government, indeed, has become the major revenue source for OldKids. For instance, in 2003, the Shanghai government organized two major events—the "One Million Families Online Campaign" (OMFOC) and the "Internet Surfing Project for Seniors" (ISPS)—to promote the adoption of the Internet in Shanghai among disadvantaged social groups such as unemployed middle-aged women and older adults.* OldKids, because of its new computer training focus and the "nonprofit" image, was easily selected by the Shanghai municipal government as one of the main contractors to provide training services during both citywide Internet-promoting events. Doing this kind of contracting work for the Shanghai municipal government generated approximately two-thirds of the annual revenue for OldKids. It also helped to promote and reinforce the nonprofit image among OldKids members.

Another factor that may have contributed to these older Chinese persons' utopian perception of the community is that the three cofounders of OldKids are widely perceived by members as having a "spirit of dedication" (*Feng Xian Jing Shen*) or "spirit of selflessness" (*Wu Si Jing Shen*). In fact, during the interviews many participants repeatedly mentioned that such a spirit of the three cofounders is an important

---

* In March 2003, four branches of the Shanghai municipal government—the Women's Association, Information Technology Office, Office of Cultural Affairs, and Science and Technology Association—initiated the OMFOC. The primary goal of this three-year campaign is to, between 2003–2005, train one million Shanghai residents, especially women in the age range of thirty-five to sixty who are unemployed, laid-off, or retired, to use computers and the Internet. The municipal government hopes that, by educating one person per household about computers and the Internet, this campaign can influence one million Shanghai families and therefore improve the overall use and adoption of computers and the Internet in Shanghai. Several months later, the Shanghai Seniors Working Committee Office, Science and Technology Association, and Seniors Foundation started the ISPS. This three-year project has three main objectives: (1) to train 100,000 older adults age sixty and above to learn and use computers and the Internet, (2) establish one hundred training bases for seniors (each training base has at least twenty computers with Internet connections), and (3) establish one thousand community centers for older adults to surf the Internet (each center has at least three networked computers). According to Wu Hanzhang, Chairman and cofounder of OldKids, and also Sun Pengbiao, director of the ISPS, the main idea of the ISPS—i.e., to help seniors get online—was originally proposed by OldKids and later on adopted and promoted by the Shanghai government as a governmental project.

reason for the formation of the OldKids *Shi Wai Tao Yuan*, because the three of them have shown the rest a "good role model" to follow (see the quotations above.)

Ironically, however, the cofounders of OldKids have their own agenda, which is exactly what the members thought they were able to avoid in the OldKids *Shi Wai Tao Yuan*: to pursue monetary gain for personal interests (instead of selflessly dedicating themselves to help their seniors.) During the interviews, the three cofounders of OldKids did not attempt in any way to hide their true intention. They all frankly admit that their primary goal is to make a profit. For instance, one of the cofounders explicitly stated:

> Of course we want to make a profit. As a private, for-profit organization, we want to and have to make money. That's the only reason why we have been doing all these since the very beginning. The seniors' market has great potential. Especially in big cities like Shanghai, seniors' retirement pensions are usually pretty good. An ordinary retired blue-collar worker normally has about RMB 1,000 per month. For retired intellectuals and cadres, their monthly pension would be even higher, around RMB 2,000. There are more than two million seniors in Shanghai …This is a really promising market; it has great potential. We hope that we can make a profit from this big market … [Wu Hanzhang, Chairman and cofounder of OldKids]

This view as described by Wu Hanzhang here is clearly market/money driven. In other words, it clearly shows that the activities that OldKids has been engaged in are no different from any other types of market activity. Why, then, would most OldKids members believe that the cofounders were just being selflessly dedicated to serving the seniors?

One possible reason is that so far OldKids has not charged much for their computer training service, even though this is because the cofounders have realized that it is the only way to attract more students to sign up for the computer classes and, in turn, to expand the population of older Internet users. To say it slightly differently, the cofounders have come to realize that, in order to make a profit from the seniors market, it is necessary to first nurture and expand the market by helping more seniors use the Internet, as the following quotation illustrates:

> We have found that seniors' conceptions of consumption are quite problematic—if you charge them too much, they won't come, even if they have no problem affording the fee. You know how stingy our seniors are. They will only come if the price is really inexpensive. So we have lowered our price, because we want to first help more seniors use the Internet. We don't plan to make money from the training at the present time. Our current goal is to nurture and expand the market so that we'll be able to make a profit in the near future, when there are more seniors using the Internet … [Wu Hanzhang, Chairman and cofounder of OldKids]

In short, OldKids provides inexpensive service at the present time because it is the only way the company can survive and also because the managers believe doing so for now will help them profit in the future. Thus, the real motivation for the OldKids cofounders to provide inexpensive computer training services for seniors is directly market driven, and selflessness or dedication has never been the real motivation behind their behaviors.

It is worth mentioning that a few OldKids members have gradually become aware of the market-driven motivation of OldKids. One OldKids member who also participates in another senior-oriented online community—SilverHair (*Yin Fa*), comments on the differences between these two online communities:

> SilverHair is created and maintained by seniors, and OldKids is managed by young people. The former is quietly but practically doing good things for seniors for free, while the latter has more emphasis on money. OldKids likes to keep a high profile for itself. [Mr. D]

Another OldKids member expresses even stronger and more explicit criticism of OldKids' market-driven services. He explicitly called the OldKids organization "overcommercialized" and pointed out that "nothing is free" here and that the cofounders "just want to make money." [Mr. Z]

Although a few members have begun to realize the market-driven trajectory of OldKids, the majority of members perceive OldKids as a *Shi Wai Tao Yuan* that is "unpolluted" by the market system. Such a perception might be a result of OldKids' effort to actively promote itself as an organization that provides not-for-profit "community service," an image that is backed up by the current low price of its service. However, this perception held by most OldKids members might also be an indication of older Chinese persons' desperation in finding allies—especially among the younger generations—who are willing to share their old values and beliefs. Disappointed and marginalized by the market system and its accompanying new values and beliefs that have come to dominate Chinese society in the past two and half decades, older Chinese people are struggling to find a refuge where their values and beliefs can still mean something to both themselves and those around them. Therefore, they are willing to accept the image promoted by OldKids, as long as it holds certain truth at the present time.

11.5 Beliefs, Norms, and Resistance

Regardless of the precise reasons behind these OldKids members' perceptions (and whether or not these reasons are justified), the majority of OldKids members' shared view of the OldKids community as a Utopia may well reflect sharp conflicts between two socioeconomic systems. In contemporary market-driven Chinese society, material enjoyment and personal interests are encouraged; in comparison, in the old economic and social systems that were in place when these older Chinese people were growing up, the emphasis was on spiritual pursuits and public interests (Jiang and Ashley 2000). The older Chinese generation feels confused and frustrated in the reform era, when the values and beliefs they had held so dearly now appear to be outdated and even scorned. The Internet has become a beacon for these older Chinese people because its spirit is largely coincident with that of these older Chinese people, e.g., equality, devotion, and sharing valuable things (like software and information) for free. In this sense, these older Chinese Internet users are creatively using the Internet to resist market-oriented changes and to restore and maintain the values and identities acquired during their youth. To use Ryff's (1989) terms, these older Chinese people have shown signs of being autonomous individuals because each one of them is "self-determining and independent, able to resist social pressures to think and act in certain ways, regulates behavior from within, evaluates self by personal standards" (p. 1072).

Similarly, based on online ethnographic fieldwork conducted in 1998, Christensen (2003) reported in *Inuit in Cyberspace: Embedding Offline Identities Online* that Inuit cultural identities have been extended and preserved in the online world. He wrote: "Instead of disembedding computer interaction from offline life, Inuit are generally embedding offline life into cyberspace, creating and asserting continuity in their lives—building on a continuum that has a strong influence in their worldview conceptions without necessarily polarizing reality into extremes of what is real or not" (Christensen 2003, p. 18). Christensen's findings among Inuit are similar to what I have found among the older Chinese participants, who are also using the Internet to continue and preserve their identities. Note that both groups—Inuit and older Chinese people—are marginalized (socially, politically, economically, and especially technologically), and yet, once they have learned the technology, they are using it as a new approach to continue, rather than discontinue, their old traditions and identities. These findings provide empirical evidence that the online world or how the Internet is used and perceived is indeed affected by the offline world—more specifically, offline factors such as identities that are formed and developed in the physical world prior to the introduction of the technology.

In the edited volume, *Chinese Society, Change, Conflict and Resistance*, the contributors addressed reform, resistance, and protest in various topical areas in contemporary China (Perry and Selden 2003). Although the contributors did not address issues related to information technology and aging, their discussions of "*resistance*"—more precisely, "*everyday resistance*"—is of particular relevance to the discussion here. Perry and Selden (2003) point out that "much everyday resistance is invisible" and that this type of resistance "takes such forms as private acts of evasion, flight and foot dragging, which, in the absence of manifestos or marches, may nevertheless effectively enlarge the terrain of social rights" (p. 2). Following this argument, it appears that the majority of older Chinese participants' view of the OldKids Internet community as a Utopia is a new form of *resistance*—that is, everyday resistance to the market system and changes that are inevitably associated with and caused by the transition from a planned economy to a market one. Compared with the modes of resistance discussed by Perry, Selden, and their contributors (2003; e.g., private acts of evasion, flight, and foot dragging), however, this form of resistance as found among these OldKids members is more mild and subtle, and, consequently, even less visible.

Older Chinese persons' resistance to the newly established market economy systems is not unique to the Chinese context. Rather, it is well documented that citizens of former socialist nations—including the former Soviet Union and Eastern European nations that have also gone through dramatic transformations in a short period of time—miss the kinds of job security, medical benefits, housing, education, and other elements of social welfare as well as the warm social and personal relations and other values that they were used to and had enjoyed before the transformations (for a review, see Tang and Parish 2000).

Farquhar and Zhang (2005), based on their ethnographic study of older residents of Beijing, argued that "The aspects of Maoist life that people of this age remember positively are the orderly parts, and the aspects of modern life they dislike are those that buy progress at the expense of order." This study did not find explicit evidence that could support Farquhar and Zhang's argument about how older Chinese people were nostalgic about "the orderly parts" or stability that was featured in their youth. However, it found that equality, devotion, and free sharing appeared to be the main aspects/values that OldKids members missed the most, contributing to their dislike of the current market system.

An important question to consider is whether the OldKids Internet community as a Utopia is an illusion or reality. On the one hand, it might be an illusion because the cofounders as well as other employees of OldKids clearly have their market-oriented agenda. On the other hand, however, OldKids members themselves have indeed been demonstrating their dedication and selflessness while interacting with one another online or offline. As such, there appears to be a solid ground for OldKids members' perception of the OldKids Internet community as *Shi Wai Tao Yuan*. Thus, this analysis does not suggest that the *Shi Wai Tao Yuan* perception that the majority of OldKids members hold is a total illusion that may not last long. Interestingly, this utopian view of the Internet community is only found in the OldKids case study—in the case study of a primarily U.S.-based online community, SeniorNet, members do not have similar perceptions of the SeniorNet community, even though they have also selflessly dedicated themselves to helping one another (Xie 2006a, 2007c). This further suggests the influence of the Chinese context in these older Chinese persons' perceptions, beliefs, and norms.

Acknowledgments

This material is based upon work supported by the National Science Foundation under Grant No. 0431373. The author would like to thank Kim Fortun for her encouragement in exploring related issues and Ken Fleischmann for his editorial assistance on earlier versions of this manuscript.

References

Bian, Y. 2002. Chinese social stratification and social mobility. *Annual Review of Sociology* 28:91–116.

Central Intelligence Agency. 2007. The world factbook. Washington, DC: Central Intelligence Agency. https://www.cia.gov/library/publications/the-world-factbook/index.html (accessed November 16, 2007).

Chappell, N. L. 2003. Correcting cross-cultural stereotypes: Aging in Shanghai and Canada. *Journal of Cross-Cultural Gerontology* 18 (2):127–147.

Chau, R., and S. Yu. 2001. Making welfare subordinate to market activities: Reconstructing social security in Hong Kong and Mainland China. *European Journal of Social Work* 4 (3):291–302.

Chen, K., and X. Cheng. 1999. Comment on Zhou and Hou: A negative life event with positive consequences? *American Sociological Review* 64 (1):37–40.

China Internet Network Information Center. 1997. The First survey report on Internet development in China. Beijing, China. http://www.cnnic.org.cn/download/2003/10/13/93603.pdf (accessed October 1, 2008).

China Internet Network Information Center. 2008. The 22th survey report on Internet Development in China (July 2008). Beijing: China: Internet Network Information Center. http://www.cnnic.org.cn/uploadfiles/pdf/2008/7/23/170516.pdf (accessed October 16, 2008).

Christensen, N. B. 2003. *Inuit in Cyberspace: Embedding Offline Identities Online.* Denmark: University of Copenhagen, Museum Tusculanum Press.

Cowgill, D. O., and L. Holmes, eds. 1972. *Aging and Modernization.* New York: Appleton-Century Crofts.

Farquhar, J., and Q. Zhang. 2005. Biopolitical Beijing: Pleasure, sovereignty, and self-cultivation in China's capital. *Cultural Anthropology* 20 (3):303–327.

Freund, E. M. 2001. Fizz, forth, flat: The challenge of converting China's SOEs into shareholding corporations. *Policy Studies Review* 18 (1):96–111.

Fry, C. L. 1999. Anthropological theories of age and aging. In *Handbook of Theories of Aging,* edited by V. L. Bengtson and K. W. Schaie. New York: Springer Publishing Company.

Glaser, B. G., and A. L. Strauss. 1967. *The Discovery of Grounded Theory: Strategies for Qualitative Research.* Chicago: Aldine, pp. 217–286.

Ikels, C. 1996. *The Return of the God of Wealth: The Transition to a Market Economy in Urban China.* Stanford, CA: Stanford University Press.

Jiang, Y., and D. Ashley, eds. 2000. Mao's Children in the New China: Voices from the Red Guard Generation. In *Asia's Transformations,* edited by M. Selden. London, UK: Routledge.

Joseph, A. E., and D. R. Phillips. 1999. Ageing in rural China: Impacts of increasing diversity in family and community resources. *Journal of Cross-Cultural Gerontology* 14:153–168.

Kazmer, M. M., and B. Xie. 2008. Qualitative interviewing in Internet studies: Playing with the media, playing with the method. *Information, Communication and Society* 11 (2):115–136.

Lee, C. K. 1998. *Gender and the South China Miracle: Two Worlds of Factory Women.* Berkeley, CA: University of California Press.

Lee, L. 2004. The current state of public health in China. *Annual Review of Public Health* 25:327–339.

Lee, P. N.-S. 2001. Restructuring China's welfare regime: A case study at the local level. *Policy Studies Review* 18 (1):59–74.

Li, H., and M. B. Tracy. 1999. Family support, financial needs, and health care needs of rural elderly in China: A field study. *Journal of Cross-Cultural Gerontology* 14:357–371.

Liu, R. 1994. *Baseline Survey Data of Beijing Multidimensional Longitudinal Study of Aging.* Beijing, China: Weijin Publishing House.

Lu, L., and I. Alon. 2004. Analysis of the changing trends in attitudes and values of the Chinese: The case of Shanghai's young and educated. *Journal of International and Area Studies* 11 (2):67–88.

National Bureau of Statistics of the People's Republic of China. 2001. Report on the 5th census. Beijing, China: National Bureau of Statistics of the People's Republic of China.

Ng, A. C. Y., D. R. Phillips, and W. K. Lee. 2002. Persistence and challenges to filial piety and informal support of older persons in a modern Chinese society: A case study in Tuen Mun, Hong Kong. *Journal of Aging Studies* 16:135–153.

Palmore, E. B., and D. Maeda. 1985. *The Honorable Elders Revisited.* Durham, NC: Duke University Press.

Palmore, E. B., and K. Manton. 1974. Modernization and the status of the aged: International correlations. *Journal of Gerontologist* 29:205–210.

Perry, E. J., and M. Selden, eds. 2003. *Chinese Society, Change, Conflict and Resistance. Asia's Transformations,* 2nd ed., edited by M. Selden. London, UK: RoutledgeCurzon.

Price, R. H., and L. Fang. 2002. Unemployed Chinese workers: The survivors, the worried young and the discouraged old. *International Journal of Human Resource Management* 13 (3):416–430.

Rook, K. S. 1987. Social support versus companionship: Effects on life stress, loneliness, and evaluations by others. *Journal of Personality and Social Psychology* 52 (6):1132–1147.

Rook, K. S. 1995. Support, companionship, and control in older adults' social networks: Implications for well-being. In *Handbook of Communication and Aging Research,* edited by J. F. Nussbaum and J. Coupland. Mahwah, NJ: Lawrence Erlbaum. pp. 437–463.

Ryff, C. D. 1989. Happiness is everything, or is it? Explorations on the meaning of psychological well-being. *Journal of Personality and Social Psychology* 57 (6):1069–1081.

Shanghai Municipal Population and Family Planning Commission. 2003. Year 2002 Shanghai population and family planning report. Shanghai, China. http://www.pop-info.govern/popinfo/pop-docrfix.usf/v.tjz1/E00B5.Et9A8E2A4825600F0u2A4605 Accessed August 2nd, 2009.

Shanghai Research Center on Aging. 2005. Year 2004 Shanghai older population statistics. Shanghai, China. http://www.shrca.org.c/1940. html. Accessed August 2nd, 2009.

Simmons, L. W. 1945. *The Role of the Aged in Primitive Societies.* New Haven, CT: Yale University Press.

Strauss, A. L., and J. Corbin. 1998. *Basics of Qualitative Research: Techniques and Procedures for Developing Grounded Theory,* 2nd ed. Thousand Oaks, CA: Sage.

Streib, G. F. 1987. Old age in sociocultural context: China and the United States. *Journal of Aging Studies* 1 (2):95–112.

Sun, R. 2002. Old age support in contemporary urban China from both parents' and children's perspectives. *Research on Aging* 24 (3):337–359.

Tang, W., and W. L. Parish. 2000. *Chinese Urban Life under Reform: The Changing Social Contract.* New York: Cambridge University Press.

Tu, E. J.-C., J. Liang, and S. Li. 1989. Mortality decline and Chinese family structure: Implications for old age support. *Journal of Gerontology: Social Sciences* 44 (4):157–168.

Walder, A. G. 1989. Social change in post-revolution China. *Annual Review of Sociology* 15:405–424.

Warner, M. 2001. The new Chinese worker and the challenge of globalization: An overview. *International Journal of Human Resource Management* 12 (1):134–141.

Whyte, M. K., E. F. Vogel, and W. L. Parish Jr. 1977. Social structure of world regions: Mainland China. *Annual Review of Sociology* 3:179–207.

Xie, B. 2005. Getting older adults online: The experiences of SeniorNet (USA) and OldKids (China). In *Young Technologies in Old Hands—An International View on Senior Citizens' Utilization of ICT*, edited by B. Jaeger. Copenhagen, Denmark: DJOF Publishing.

Xie, B. 2006a. Growing older in the information age: Civic engagement, social relationships, and well-being of older Internet Users in China and the United States. PhD diss., Rensselaer Polytechnic Institute, Troy, NY, pp. 175–204.

Xie, B. 2006b. Perceptions of computer learning among older Americans and older Chinese. *First Monday* 11 (10). http://firstmonday.org/htbin/cgiwrap/bin/ojs/index-php/fm/article/view/1408/ Accessed August 2nd, 2009.

Xie, B. 2007a. Information technology education for older adults as a continuing peer-learning process: A Chinese case study. *Educational Gerontology* 33 (5):429–450.

Xie, B. 2007b. Older Chinese, the Internet, and well-being. *Care Management Journals: Journal of Long Term Home Health Care* 8 (1):33–38.

Xie, B. 2007c. Using the Internet for offline relationship formation. *Social Science Computer Review* 25 (3):396–404.

Xie, B. 2008a. Civic engagement among older Chinese Internet users. *Journal of Applied Gerontology* 27 (4):424–445.

Xie, B. 2008b. Multimodal computer-mediated communication and social support among older Chinese. *Journal of Computer-Mediated Communication* 13 (3):728–750.

Xie, B. 2008c. The mutual shaping of online and offline social relationships. *Information Research* 13 (3): paper 350. http://InformationR.net/ir/13–3/paper350.html. Accessed August 2nd, 2009.

Xie, B., and P. T. Jaeger. 2008. Older adults and political participation on the Internet: A cross-cultural comparison of the United States and China. Journal of Cross-Cultural Gerontology 23:1–15.

Zhou, X., and L. Hou. 1999. Children of the Cultural Revolution: The state and the life course in the People's Republic of China. American Sociological Review 64 (1):12–36.

Zhou, X., N. Tuma, and P. Moen. 1997. Institutional change and job shift patterns in urban China. American Sociological Review 62 (3):339–365.

Chapter 12

Social Network Sites
An Exploration of Features and Diversity

Mike Thelwall and David Stuart

Contents

12.1 Introduction

MySpace™ and Facebook™ have been two of the biggest commercial successes of the past decade. Created in 2003, by July 2008 they were the seventh and fifth most visited Web sites in the world, respectively, according to Alexa™.* Although having primarily English-speaking members, similar social networking sites have been successful around the globe, including Brazil (Orkut™, http://www.orkut.com), South Korea (Cyworld™, http://www.cyworld.com), Japan (Mixi™, http://mixi.jp), Germany (StudiVZ™, http://www.studivz.net), France (Skyrock™, http://www.skyrock.com), Iran (Cloob™, http://www.cloob.com), and Russia (V Kontakte™, http://vkontakte.ru). Social networking seems to be a global phenomenon, successful everywhere having wide Internet access (excluding most of Africa.) As such, it seems to have tapped into a universal and fundamental human desire—the need to be sociable. Nevertheless, this phenomenon is so new that it is not yet well understood. How do these sites embed within members' lives? What does friendship mean online?

This chapter gives an introduction to current research in order to reveal insights into the range of features offered by social network sites (SNSs) and the diversity of their uses. There are many different successful SNSs, each with its own unique collection of features and its own unique types of users. But are the sites essentially the same or fundamentally different? And how important are cultural differences in usage? Exploring the features of diverse SNSs provides a useful way to understand their possibilities and how very different approaches have managed to generate successful and unique SNSs.

This introduction continues with a definition and broad typology for SNSs. This is followed by a description of seven diverse SNSs, giving a brief outline of each and picking out one or more significant features of each site in order to explore a particular issue. The key concept of friendship is the first such feature explored, but other features include gift giving, photograph sharing, and micropayments. A range of relevant issues are also discussed, again in relation to specific SNSs; these include security and identity expression. The nature of social networks is then discussed as they increasingly reposition themselves as application platforms, while users demand greater access to the data they create. Finally, the conclusion incorporates a table of key issues and some comments on how these differ between sites. By the end of this chapter, the reader who is perhaps familiar with only one or two SNSs should have

---

* http://www.alexa.com/topsites.

a much broader understanding of their variety and international spread, as well as a background understanding of some of the interesting social issues relating to them.

12.1.1 Definition

An SNS is a Web site that allows people to join, own, and edit a personal profile page, to publicly connect to other members (e.g., as "friends"), and to communicate with other members (Boyd and Ellison 2007). This definition matches the core sites like Facebook and MySpace but is broad enough to include sites like YouTube™, which most users would see as a video-sharing site rather than an SNS. In this chapter, however, the main focus is more specific: SNSs that are primarily for socializing among friends and acquaintances, as elaborated below. The dropping of the "ing" from "network" in the SNS definition is controversial (Beer 2008). On one hand, it recognizes that the biggest SNSs are not for network*ing* in the sense of making new friends or business contacts, but on the other hand, the term "networking" also emphasizes the interpersonal communication aspect, which is often important (see below).

As the definition above highlights, public friendship is the key SNS component. Before SNSs, a person's online friends might have been recorded in private places like their e-mail contact list, their bookmarks to friends' home pages, and their memory of the chat room nicknames of frequent communicators. None of these are proto-SNSs because the friend lists are not public. Public friend lists (typically including a proportion of genuine friends) are important because they can be navigated to find friends of friends. When a new user joins an SNS, their first action may be to friend one offline friend and then browse their friend list to discover and friend mutual friends and acquaintances.

The personal profile aspect of the SNS definition—apparently ubiquitous in sites with a friending mechanism—reflects identity information (real or fake) that helps make judgments about who to friend or gives more information about friends. This information tends to include standard personal details, such as age, gender, and geographic location, but Mixi also includes blood type, because this is important in Japanese culture (Komaki, preprint), and role-playing sites like Gaia Online™, (www.gaiaonline.com) may give information about the member's role rather than her offline identity.

12.1.2 Typology

As mentioned above, some Web sites fitting the SNS definition are not primarily for social networking or for communication within existing social networks. There seem to be three alternative types of SNS, as shown in Figure 12.1. *Socializing SNSs* support informal social interaction between members. This type includes Facebook and MySpace and probably reflects most people's idea of a SNS. *Networking SNSs* support nonsocial interpersonal communication, with the business networking sites LinkedIn™ and Ryze™ being prime examples. Friendster™'s early dating focus also

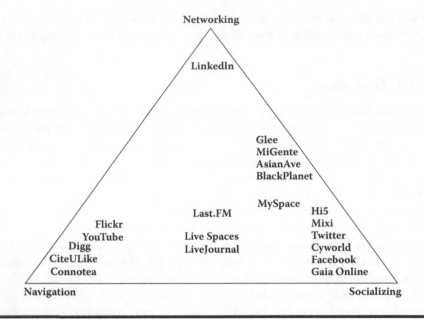

Figure 12.1 Typology Triangle (From Thelwall, M. *Online Information Review* 33(1): 587. Used with permission.)

classed it as a networking SNS, as its friendship lists seemed to be designed for members to find dates by surfing friends of friends (Boyd 2004). Networking SNSs are essentially people-finding SNSs.

In contrast, (Social) *Navigation SNSs* support finding resources via interpersonal connections. These resources could be photographs (Flickr™), videos (YouTube), blogs (LiveJournal™, Live Spaces™), or art (DeviantArt™) within the SNS. They could also be Web pages (delicious™, StumbleUpon™), news (Digg™), or academic references (CiteULike™, Connotea™) stored outside of the SNS. The role of SNS friendship here is primarily social navigation: to help users find relevant content via finding relevant people rather than to engage in social interpersonal communication. For example a Linux-using Digg member might subscribe to Digg members who are good at finding Linux-related online news. Friendship in this type of SNS is often asymmetrical: Permission is not needed to become someone's friend/follower/subscriber because the purpose of the connection is to give easier access to the creations of the targeted member.

Many SNSs are powerful enough to be used for all three functions, even if their primary use or design is more specific. For instance, MySpace allows members to post their own pictures and videos, and links can be added to personal profiles, blog postings,

---

* Sites are classified based on an estimate of the degree to which they use each of the three extremes.

or comments. As a result, the triangle typology illustrated in Figure 12.1 is based on the apparent intended or actual use of a site rather than its technical functionality.

12.2 MySpace™: Friendship, Identity, and Security

MySpace was the second most visited social SNS in the world in July 2008 according to Alexa,* having fallen behind Facebook that year. Out of a list of all types of Web sites, in the United States it was ranked third overall for visitors, in the United Kingdom it was eleventh, but in some other countries it is far less significant; for example, it is ranked twenty-seventh in South Korea, forty-eighth in Brazil, ninety-fourth in Japan, and 128th in China. Overall, then, in 2008 MySpace was one of the top two SNSs but was far from globally dominant.

MySpace was created in 2003 in order to fill a gap in the provision of the top SNS at the time, Friendster (Boyd 2006; Boyd and Ellison 2007). Friendster was intended for friendship circles and perhaps also acquaintances, but some users were competitively trying to attract huge numbers of "friends," and bands were using it to promote their work among fans and potential fans. Friendster was oriented toward dating, and so the company restricted fake profiles, even though many were highly popular (Boyd and Ellison 2007). In contrast, and perhaps to partly fill a perceived gap between Friendster features and its members' aspirations, MySpace was designed from the start to allow bands to promote their music. Presumably as a result of this, it attracted a particular following among youth as it rapidly grew.

MySpace members have a profile page containing a photo of themselves, some personal information, a song playing, lists of personal interests, a list of friends, and a list of friends' comments. In addition, there are links to information on other pages, such as blog postings, videos, and pictures. Friending is reciprocal in MySpace so that to become friends with someone, members must find them in MySpace (perhaps by surfing lists of friends of existing friends) and send them an official friendship request. The recipient must confirm the request in order to register the relationship. MySpace profile pages have a standard look, but members can override this with their own background images and color schemes. Friends can communicate either by sending private messages or by writing on each other's public comments area. Members can also broadcast to all friends via their blog, pictures, and videos.

Although much media attention and research have focused on teenage members, these seem to account for less than half. The average declared age of MySpace members was twenty-one in 2007 (Parks 2008; Thelwall 2008b), which is still relatively young. It seems likely that the average age of MySpace members will increase as its existing members age and are balanced by new, younger members; although it is also possible that users will migrate to other SNSs that they perceive to be more

---

* http://www.alexa.com/data/details/traffic_details/myspace.com?site0=www.myspace.com, accessed July 30, 2008.

appropriate as they age. There is an approximately even split between males and females and young U.S. members seem to be, on average, less educated than those of Facebook (Boyd 2007).

12.2.1 Friendship

The interpretation of friendship is central to socializing SNS members' activities. It is important to recognize that the notion of social network friendship is typically weaker than offline friendship (Boyd 2004). There seem to be three different common meanings for MySpace friends: "real" friends, acquaintances, and nothing. For many people, the MySpace friends that they interact with most are their offline friends (Boyd 2006) and sometimes their registered MySpace friends may be just these. At the other extreme, musicians and some individuals collect as many friends as possible without any intention of initiating a dialogue or other natural friendship activity. This could be classed as a fan or casual relationship with no real friendship component. In between the two extremes, people may initiate and accept friendship requests from acquaintances who they would not view as "real friends." Reasons for initiating or accepting such friendship requests include the following (for Friendster and MySpace, but all are probably true for MySpace):

1. They are acquaintances, family members, or colleagues.
2. It would be socially inappropriate to say no if you know them.
3. Having lots of friends makes you look popular.
4. It's a way of indicating that you are a fan (of that person, band, product, etc.).
5. Your list of friends reveals who you are.
6. Their profile is cool so being friends makes you look cool.
7. Collecting friends lets you see more people (Friendster.)
8. It's the only way to see a private profile (MySpace.)
9. Being friends lets you see someone's bulletins and their friends-only blog posts (MySpace.)
10. You want them to see your bulletins, private Profile, or private blog (MySpace.)
11. You can use your friends list to find someone later.
12. It's easier to say yes than no.

(Boyd 2006)

Some MySpace research suggests that lists of two to nine friends are predominantly close friends, ten to ninety are predominantly acquaintances, and ninety-one or more are predominantly strangers (Thelwall 2008b). The average (median) number of MySpace friends for those with at least two is twenty-seven. Probably, however, most members have some real friends, some acquaintances, and some strangers—probably musicians—but communicate mainly with offline friends.

MySpace is a site in which not all friends are necessarily equal. A member can choose which friends to include within their top friends list, as displayed on the home page. Such lists presumably tend to include close friends and perhaps also favorite bands. Among young U.S. members, this feature has caused some social problems because of the highly visible and clear-cut nature of the distinction (Boyd 2006). In practice, it may be socially desirable to let old friendships drift rather than to publicly announce that someone else has taken over as a best friend by adjusting the top friends list. The same rationale probably helps to explain why members choose to keep a loose working definition of friendship within SNSs.

12.2.2 Identity Expression

MySpace often seems to be a place of identity projection and personal expression, through the appearance of the profile, the people in the friend lists, and the content of comments written on profile pages (Boyd 2007). Some members have even felt that MySpace provides a forum within which their skills can be appreciated among their peers (e.g., witty comments) in a way that was not possible offline (Boyd 2007).

The creativity of the visual appearance of many MySpace profiles is an attractive feature. Profile customization is not easy in MySpace but is possible by inserting the appropriate hypertext markup language (HTML) code into profile information fields, effectively hacking the page. Hacking HTML ought to be beyond the capability of most MySpace users but they circumvent the technicalities by copying and pasting from friends or special sites (Perkel 2006). As a result, teen MySpace sites are often very visual reflections of the identity that they have tried to project to their friends. Visual customization is far from ubiquitous in SNSs, with Facebook and many other sites not supporting it. In contrast, sites like Cyworld and Gaia Online that have a virtual world aspect encourage members to customize this virtual world, while allowing the overall profile to be customized to some extent too.

Some research suggests that younger teenagers tend to favor visual customization for identity projection, whereas older teenagers, and presumably adults, prefer to use their friend lists for this purpose (Livingstone 2008). In other words, older members seek to project an identity through the type of people who they have friend connections to. Because of the presence of friends' comments in profile pages, others can also help in the formation of the identity of the profile owner, particularly through positive or negative comments (Walther et al. 2008).

12.2.3 Comments

A common way for members to communicate in social SNSs is by leaving public comments on each other's profile. These comments seem to be typically part of a two-way dialogue, but because they are public and listed with other friends' comments, it is likely that they will be read by many of the friends of the comment recipient. As a result the comments may contain more of a performance element than private communication.

Some studies have analyzed the language of MySpace comments, finding it to be typically very informal, with a majority of comments not being written completely in standard formal English (Thelwall 2009a), and with swearing being present in a majority of profiles (but probably mainly in the comments) and particularly common for younger members (Thelwall 2008a). The presence of swearing emphasizes the informal nature of this kind of communication and perhaps also indicates a desire to inject emotion into messages.

12.2.4 Security

One important issue for the press in the early years of MySpace was the security of its young members, especially safety from pedophiles (e.g., Rawsthorne 2006). This concern led to MySpace setting a minimum age of fourteen and hiding profile information from strangers for users younger than sixteen. Older users can also restrict access to their profile information to friends, but only about 18% do—mostly females (Thelwall 2008a). Teen users predominantly keep sensitive information out of their profiles (e.g., phone, address), but a minority are apparently careless enough to refer to illegal activities such as drug taking and underage drinking (Hinduja and Patchin 2008). The majority of U.S. teen social network users believe that a persistent visitor can find out who they are from their profile information alone, which suggests a degree of acceptance of risk or a disregarding of the problem (Lenhart and Madden 2007). Perhaps in response to such slips online, MySpace has taken steps to remove identified pedophiles (MSNBC 2007), although the majority of these were probably United States based, and even then there is no indication of how successful the process was.

There is no simple answer to whether MySpace has made life more or less safe for teenagers, although it is clear that the new social environment requires new social skills. U.S. teens have used MySpace to hang out with their friends, substituting cyberspace for the mall that their older siblings may have been allowed to visit (Boyd 2007), and as such, MySpace seems to have provided a relatively safe communication environment to a new generation (see also Ybarra and Mitchell 2008). However, the potential dangers faced in a public mall are more obvious than those faced by a teenager in the perceived security of his own bedroom, and it is not clear whether MySpace's™ publically removing identified pedophiles will create a false sense of security or raise awareness of the need to be careful.

A different security issue is the ability of strangers, including potential employers and the police, to view profile pages and to use this information for important decisions. Although there have been stories in the press about people being rejected for jobs on the basis of unsuitable SNS content, this does not seem to be a concern for many people. For example, a study of students found most to be unconcerned about this issue (Tufekci 2008a). SNS members have to make judgments about how much information to reveal about themselves, however, especially if they intend to make or cement friendships online because this process typically involves various forms of information disclosure (Dwyer et al. 2007).

12.3 Cyworld™: Micropayments and Family Relationships

Cyworld is a South Korean site that is the world's first mass SNS. It added SNS functionality to its previous incarnation as a blog-based site in 2001, subsequently growing to near ubiquity among South Korean Internet users. Like MySpace, it is primarily designed for social communication but, unlike MySpace, does not emphasize music. Cyworld users have a profile page (mini-home page, or mini-hompy) but also "exist" as an avatar (minime) in Cyworld and have their own virtual home displayed prominently in their minihompy, as shown in Figure 12.2. This home can be furnished and decorated to the owner's taste, providing an outlet for creative expression similar to MySpace profile customization, although this customization is not free in Cyworld (Haddon and Kim 2007). A minihompy visitor may see

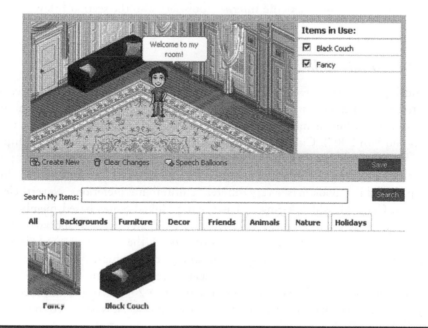

Figure 12.2 A Cyworld mini-room from a member's profile page showing part of the editing environment.

the owner's avatar displayed living in their personalized home, perhaps with some visiting friends. As a result of this, Cyworld is clearly branded as an online alternative world. Despite this, uploading photographs and keeping a diary are important activities in Cyworld, both representing very direct connections to the owner's offline existence (Haddon and Kim 2007).

Although no age demographics are available for Cyworld, its high rate of use in South Korea means that it must be popular among people of most ages, if not all. It is particularly relevant (even for the future of MySpace) that its membership is not restricted to youth despite its playful feel and cartoon graphics. Security problems seem minimal with Cyworld in South Korea because new users' identities have to be verified through their government identification, and privacy issues are dealt with through a system that allows different types of people access to different aspects of the site. This is different in the newer, U.S. version of Cyworld, however, which does not require proof of identity.

In 2008, Cyworld appeared to have reached saturation point in South Korea, and its owners judged that its SNS functionality might not be enough to sustain its interest. Their reaction was to attempt to turn Cyworld into a portal, for example by adding search engine functionality prominently to the home page (Jin-seo 2008). Presumably the logic was that this would provide another reason for members to keep visiting the home page and hence a constant reminder to log into Cyworld. This seems to be a possibility for the future of other SNSs too—becoming integrated into a range of other core Internet tools. See also the section below, "SNSs as a Platform."

12.3.1 Family Relationships

Friendship (*ilchon*) in Cyworld is ostensibly not the same as in other SNSs, because this term connotes a family relationship rather than a friendship, even though, like MySpace friendship, members are free to place their own interpretation on the term (Kim and Yun 2007). Claims have been made that family-like close friendship is a peculiarly South Asian cultural phenomenon (Kim and Yun 2007), so it is possible that the technical similarity between Cyworld friends and friends in SNSs like MySpace and Facebook masks very different cultural interpretations. Nevertheless, the U.S. Cyworld, which seems to have many United States based members of Korean origin, has adopted the term "friend" rather than *ilchon* (or "special buddy," which has also been used as a translation of *ilchon*.) The success of Cyworld may be partly due to the obligation implicit in family relationships, which means that accepting an *ilchon* request may often indicate acceptance of the responsibility to regularly interact with the new *ilchon* (Kim and Yun 2007). As a result, members may be more active than those of other SNSs who may feel less obligated to interact with friends.

The family nature of friendship in Cyworld and also the popularity of purely online friendships (Choi 2006) seem to be instrumental in enabling Cyworld to perform the important social function of facilitating members to share their inner thoughts in ways that would be difficult or impossible offline (Kim and Yun 2007). Hence, despite its relatively playful appearance, Cyworld seems to fulfill a significant and useful social role and seems to be used by its members rather differently than MySpace and Facebook.

12.3.2 Micropayments

Cyworld users are expected to pay for customization, music, and mobile phone connectivity services using the Cyworld *acorn* microcurrency, which is bought with real money. This is an important source of revenue for Cyworld's owners, the Korean SK Telecom company, which gives Cyworld a business model that is different from most Western SNSs (Gaia Online is a partial exception). Many of the items bought with acorns have a limited life, but users seem to expect customization and hence give acorns as presents to ilchon to help them pay for their site (Choi 2006). As a result, maintaining a minihompy and active ilchon can be quite expensive, even if not using mobile phone access services.

Most Western SNSs seem to be financed primarily by advertising, which may be more attractive to consumers than paid services. Nevertheless, as the present giving discussed below illustrates, financial transactions can also play a useful social function. A similar example is the Facebook gifts feature, which allows members to buy virtual gifts to send to friends. The value of such gifts may be primarily due to their price— indicating an investment in friendship. LiveJournal (like Flickr) has yet another business model; it operates on a two-tier basis: free and professional. Upgrading the basic free LiveJournal gives members access to more features and bandwidth.

12.4 Gaia Online™: Role-Playing and Children

Gaia Online is a cartoon-like playful SNS aimed at young people. Like Cyworld, members have a cartoon avatar that lives in its own room that members can decorate (see Figure 12.3). The Gaia theme is games and role-playing, although chat room–like forums are also important. Gaia Online has a number of games that members are encouraged to play, either solo or interactively with other members (e.g., jigsaw, rally, a word game). Members can also customize their appearance and wander round virtual towns meeting others. This SNS seems to emphasize meeting strangers but makes this activity relatively safe for children by keeping their online identity, at least as expressed by their profiles, separate from their offline identity. The same approach is used by other child-friendly SNSs, including Club Penguin.

There is apparently no published research about Gaia Online or similar sites (as of mid-2008), perhaps due to the youth of the members. In contrast, non-SNS online role-playing sites like *World of Warcraft* seem to have attracted much more academic interest.

Figure 12.3   A Gaia Online member's home with a little furniture.

12.5 Facebook™: Pure Communication?

Facebook apparently became the most popular SNS in the world in mid-2008. It grew from a college network SNS (Harvard, 2004, all U.S. colleges, 2005) to an organizational network SNS (early 2006) until it opened up to everyone in late 2006 (Boyd and Ellison 2007). It still seems to retain college students as its core user base and in 2008 seemed to overtake MySpace in terms of total visitors. From a commercial U.S. perspective it could be more attractive to advertisers than MySpace (and hence more profitable) due to its ultimately richer, better-educated users (Boyd 2007). Friendships in Facebook seem to be predominantly between existing offline friends or acquaintances (e.g., Dwyer et al. 2007), although members seem to use Facebook often to find out about people who they have already met offline (Lampe et al. 2006), perhaps as part of a decision about how closely to befriend them. Facebook seems to help people to maintain existing friendships and to be most beneficial for people with low self-esteem (Ellison et al. 2007). As such, it seems to play a positive role in the social life of its members.

In comparison to MySpace and Cyworld, the appearance of Facebook is plain. To date (mid-2008) users cannot customize their profile appearance other than by adding applications to predefined parts of their profile. Before the introduction of applications in 2007, the key activities seemed to be personal messaging, public messaging (wall posts, similar to MySpace comments), sharing personal photographs, and joining discussion groups.

One of Facebook's distinctive features is the "Poke." Members can virtually poke each other. Poking is a content-free message because the targeted user merely receives a message telling them who poked them. Users can agree on their own meanings, however, and it seems to be a predominantly playful device.

Socializing SNSs like MySpace and Facebook seem not to attract people who do not value informal social communication like gossip (Tufekci 2008b). This suggests that social SNSs use will never be ubiquitous in society and that the heaviest users will be those who value social communication for its own sake.

12.5.1 Patterns of Use

A large-scale early study of Facebook tracked members in college networks to see how they used it (Golder et al. 2007). Facebook was not primarily used to break down geographic distance because most communication occurred between friends at the same college. Distance bridging was important during college breaks, however, when it saw an increase in messaging. Facebook use seemed to fit in with study cycles, so students probably used it when they were already studying at their computer. Facebooking did not seem to replace offline socializing because its use was low on Friday and Saturday nights—the traditional times for going out (Golder et al. 2007). This study seems to be unique in SNS research, and so we can only speculate about whether patterns of use are similar for other sites.

12.5.2 Shared Photographs

One extremely popular feature of Facebook (Joinson 2008) and Cyworld (Choi 2006) is the ability to upload pictures and share them with friends. In Cyworld, pictures can easily be edited, which adds an extra dimension. Photograph sharing is widely used in Facebook (Raacke and Bonds-Raacke 2008), perhaps because photographs can be tagged with the identities of any pictured Facebook members. This makes them more personal and allows users to quickly browse (almost) all pictures of the friend who had taken them. Photographs in SNSs can normally be commented on by friends, making them the starting point for a dialogue in many cases. Hence photographs can be an active rather than a passive element of SNSs.

12.6 LiveJournal™: Content-Based Interaction

LiveJournal is a blog-based content-sharing navigational/social SNS, although it is close to a social SNS. Users maintain their own journal, which is essentially a blog, and seem to primarily interact with each other via these journals. Friending in LiveJournal is not reciprocal: It is intended to indicate interest in reading a journal. As a result of this, the value of a person's journal can be estimated by the number of people who have chosen to friend the person and the number of comments made to the journal. LiveJournal friending can become quite political, with friendship exchanged for commenting or other services (Fono and Raynes-Goldie 2007).

LiveJournal friendship has been investigated from the perspective of member perceptions. It seems that, as in MySpace, there are some very different interpretations, and this can lead to problems. For example a person viewing LiveJournal friends as real friends may become upset if they are dropped for not updating their journal. In contrast, someone viewing friendship as merely a convenience to keep track of interesting journals might not expect to cause distress when severing a friend connection (Fono and Raynes-Goldie 2007).

12.6.1 Gifts

The financial element of LiveJournal and the relatively strong bonds between members seem to have created a culture where gift giving is not uncommon (as with Cyworld, but perhaps Facebook gifts have not taken off to the same extent.) These gifts can take the form of effort on another user's behalf or even presents with an immediate monetary value (Pearson 2007). The gift giver may be motivated by enhancing their community reputation, participating in a group, or satisfying an obvious need of the recipient; or the giver may expect eventual reciprocation (Pearson 2007). In wider society gift-giving performs a wide range of social functions (Sherry 1983), including the important and similar social functions of cementing friendships and expressing feelings, although gift giving can also be seen as fulfilling an obligation or performing a duty (e.g., Gould and Weil 1991; Komter 1997).

12.7 BlackPlanet™: Community Niche Site

Some SNSs are targeted at minority communities, apparently mainly in the United States. These include BlackPlanet™, AsianAvenue™, MiGente™ (Latinos), and Glee™ (lesbian, gay, bisexual, and transgendered people), all owned by Community Connect, Inc. (Byrne 2007). Of these, BlackPlanet seems to be the most successful and most researched. The idea underlying the three ethnic sites is to exploit the mostly racial divisions/needs in U.S. communities to deliver a targeted social environment. In MiGente this targeting includes Spanish language support, and in AsianAvenue language support is given for a range of Far Eastern languages. BlackPlanet, in contrast, does not offer any specific Black features, although Black community issues are discussed in its forums. One distinctive feature of all these community niche sites is a dating orientation, for example, with separate facilities for finding and attracting suitable dates.

There are arguments for and against community niche sites. Niche sites can be positive in that they allow minorities to seek support and operate in a positive environment, but they can also be negative in the sense of ghettoizing members away from wider society. BlackPlanet does not have a restrictive entry policy, and so the positive environment that it provides is not absolute: Racists can potentially access the site to abuse black people, although this activity is likely to result in them being expelled.

12.7.1 Discussion Forums

BlackPlanet's discussion forums cover many issues although Black community issues are a common theme. This seems to be a positive feature, but these discussions have been criticized from a political perspective for staying online and not translating into offline activities (Byrne 2007). Other environments have shown that online communication can sometimes translate into offline action (Garrido and Halavais 2003); so it seems likely that this will occasionally happen in BlackPlanet, if not in the other three sites.

Discussion groups are also popular in Facebook but are perhaps more frequently joined than actively engaged in. They are often frivolous statements rather than genuine discussions. An illustrative name is "I got through my medical degree with Wikipedia." Groups are also created for specific causes (e.g., "Un millón de Voces contra las FARC").

12.8 Last.FM™: Hybrid SNS

Last.FM™ combines some of the elements of socializing, networking, and navigation Web sites. It is based on an audio player and allows members to listen to music that they like via this player embedded in their profile page. The user's music tastes are recorded indirectly by tracking what they listen to and directly by the user rating tunes currently playing. Music can be chosen at random, played from music stored on the user's own computer, or played from Last.FM groups or other members' playlists. Each member also has a profile page with basic personal information, especially their music tastes. The profile page includes a friend list but also information about their favorite artists and tracks.

At one level, Last.FM is a music-playing service, but its strength is based upon its huge database of members' listening preferences (Audioscrobbler). While listening to a track, a user may be prompted with other tracks listened to by people who enjoyed the current track as well as the names and pictures of people who are also currently listening to it. Hence, Last.FM supports social navigation through friendship because users can find other music by friending people with similar taste. It also supports networking in the sense of finding people with similar musical taste, but because music is an important aspect of friendship, especially for youth, the connections made may well tend to be social connections. Hence Last.FM also functions as a social SNS (Baym 2008).

Other similar media-sharing or viewing sites include iLike and MOG. YouTube has ostensibly similar features but is probably used very differently in practice because the friending feature of YouTube seems to be rarely used.

12.9 SNSs as a Platform

The potential of SNSs as portals and as hybrid services has been enhanced with the introduction of application platforms by some of the major SNSs. In May 2008 Facebook opened its platform to third-party developers, enabling programmers to create applications that could be embedded in users' personal profile pages. Applications include those that reinforce Facebook's socializing aspect, as well as those that aid in navigation and networking.

The success of the Facebook platform can be seen both in the number of applications that have been developed and in the number of times these applications have been installed. According to Adonomics (2008), a provider of Facebook analytics, as of October 2008 there were 43,601 applications on Facebook, with the top application having been installed 35,135,500 times.

While the Facebook platform may be seen as a successful attempt to increase the appeal of the site and to enhance the user experience (Gjoka et al. 2008), it is also important to see it within the context of cloud computing, where information and programs are stored permanently on Internet servers rather than on personal computers (Hewitt 2008). Cloud computing is seen as increasingly important, as members of the public have increasing numbers of devices that may be used for accessing the Web in different places, using less powerful mobile devices. While various models have been proposed for the provision of services (Ballon and Walravens 2008), Facebook's establishment of a platform, which it has made available to others (Friendster 2008), has seemingly put it in a strong position.

The opening of the Facebook platform to third-party developers has been followed by similar moves by other large SNSs, including Bebo™ (Bebo 2008) and MySpace (BBC 2008). Other recent initiatives, such as Facebook Connect and Google Open Connect, are starting to allow general Web sites to integrate functionality from existing SNSs so that visitors can access features from their social network within the site.

12.10 Distributed Social Networks

With users spending a lot of time enhancing their personal profiles on SNSs, it is understandable that questions have been raised about who owns the data that are entered: whether users should be allowed to take the data they have entered on one SNS to another SNS, as well as what the SNS should be allowed to do with the data that is entered. Although this has led to the establishment of a number of projects to establish open social network protocols (e.g., DiSo Project, The Friend of a Friend Project), a realistic alternative to proprietary SNSs has yet to be established.

12.11 Summary

This chapter has introduced the three different types of SNS: social, networking, and navigational, although it has focused on social SNSs. The examples of different types of SNSs illustrate many different approaches, some of which are dominant in particular countries and some of which are more generally used (see Table 12.1). In

Table 12.1 Features of Social Network Sites

Feature	Comment
Purpose	Networking, socializing, or navigation
A profile page with basic personal information	Common to all sites
Public friending with public friends list	Common to all sites but friending not reciprocal in some sites (e.g., LiveJournal)
Costs of service	Core services always free; "professional" features require extra payment in some (e.g., LiveJournal); micropayments may support a range of additional services (Gaia Online, Cyworld)
Public commenting on friends' profiles	Common to all sites but with different names (e.g., wall posts, comments, testimonials)
Blogs	Core to some sites (e.g., LiveJournal), optional in others (e.g., MySpace, Facebook)
Photoblogs or online photographs	Popular facility in some sites (e.g., Facebook, Cyworld)
Private messaging	Common to all sites?
Instant messaging	Tends to be offered in socializing SNSs like Facebook and MySpace
Real/virtual world	Sites like Gaia Online explicitly set up a virtual world for an alternative identity
Audience	Orientation on children, youth, adults; specific languages or community groups
Groups or discussion groups	Popular in Facebook and community sites such as BlackPlanet
Gift giving	Supported in sites, including Cyworld, LiveJournal, Facebook via payments
Additional services	Music playing, videos, photo hosting, games

addition to the features offered by the sites, there is a rich user culture of actions that have not been designed into the systems, such as the use of comments, the exact nature of friendship, and gift giving. This makes SNSs complex social environments.

It will be interesting to see which of the competing SNS formats becomes dominant in the future, or if there will continue to be many different types. The relatively plain Facebook seems to be in a very strong position, and the flexibility afforded by its applications may help to keep it one step ahead of the other sites. On the other hand, the market seems driven by youth, and younger people are highly susceptible to following trends; so it may be that Facebook will become seen as old fashioned at some stage in the future. Nevertheless, it seems that social networking is so popular that it is likely to survive in some form for a considerable time.

References

Adonomics. 2008. Facebook™ analytics and developer services. Retrieved September 13, 2008, from http://adonomics.com/.

Ballon, P., and Walravens, N. 2008. Competing platform models for mobile service delivery: The importance of gatekeeper roles. Seventh International Conference on Mobile Business. Barcelona, Spain. July 7–8, 2008 pp. 102–111.

Baym, N. 2008. Tunes that bind?: Predicting friendship strength in a music-based social network. Retrieved October 21st, 2008. from http://www.onlinefandom.com/wp-content/uploads/2008/10/tunesthatbind.pdf.

BBC News. 2008. MySpace™ opens doors to developers. Retrieved October 21, 2008, from http://news.bbc.co.uk/1/hi/technology/7217975.stm.

Bebo. 2008. Bebo Apps: Open for Business. Retrieved October 21, 2008 from http://developer.bebo.com/blog/index.php/2008/01/10/bebo-apps-open-for-business/.

Beer, D. 2008. Social networking sites…revisiting the story so far: A response to danah boyd and Nicole Ellison. *Journal of Computer-Mediated Communication,* 13:516–529.

Boyd, D. 2004. Friendster and publicly articulated social networks. In Conference on Human Factors and Computing Systems CHI 2004, Vienna, Austria. April 24–29. New York: ACM Press. Retrieved July 3, 2007 from http://www.danah.org/papers/CHI2004Friendster.pdf.

Boyd, D. 2006. Friends, Friendsters, and MySpace™ Top 8: Writing community into being on social network sites. *First Monday,* 112. Retrieved June 23, 2007 from http://www.firstmonday.org/issues/issue2011_2012/boyd/index.html.

Boyd, D. 2007. Viewing American class divisions through Facebook™ and MySpace™. *Apophenia Blog Essay June 24.* Retrieved July 12, 2007 from http://www.danah.org/papers/essays/ClassDivisions.html.

Boyd, D., and Ellison, N. 2007. Social network sites: Definition, history, and scholarship. *Journal of Computer-Mediated Communication,* 13. Retrieved December 10, 2007 from http://jcmc.indiana.edu/vol2013/issue2001/boyd.ellison.html.

Byrne, D. 2007. Public discourse, community concerns, and their relationship to civic engagement: Exploring black social networking traditions on blackplanet.com. *Journal of Computer-Mediated Communication,* 31. Retrieved May 18, 2008 from http://jcmc.indiana.edu/vol2013/issue2001/byrne.html.

Choi, J. H.-j. 2006. Living in Cyworld: Contextualising Cy-Ties in South Korea. In A. Bruns and J. Jacobs, Eds., *Use of Blogs Digital Formations*. pp. 173–186. New York: Peter Lang.

Dwyer, C., Hiltz, S., and Passerini, K. 2007. Trust and privacy concern within social networking sites: A comparison of Facebook™ and MySpace™. Proceedings of AMCIS 2007, Keystone, CO. August 9–12, 2007. Retrieved May 18, 2008 from http://csis. pace.edu/~dwyer/research/DwyerAMCIS2007.pdf.

Ellison, N. B., Steinfield, C., and Lampe, C. 2007. The benefits of Facebook™ "friends": Social capital and college students' use of online social network sites. *Journal of Computer-Mediated Communication,* 12:1143–1168.

Fono, D., and Raynes-Goldie, K. 2007. Hyperfriendship and beyond: Friendship and social norms on Livejournal, Association of Internet Researchers AOIR-6, Chicago. In M. Consalvo and C. Haythornthwaite, Eds., *Internet Research Annual Volume 4: Selected Papers From the Association of Internet Researchers Conference*. New York: Peter Lang.

Friendster. 2008. Friendster supports applications from Facebook™ developer community. Retrieved October 21, 2008 from http://blog.friendster.com/2008/10/friendster-supports-applications-from-facebook™-developer-community/.

Garrido, M., and Halavais, A. 2003. Mapping networks of support for the Zapatista movement: Applying social network analysis to study contemporary social movements. In M. McCaughey and M. Ayers, Eds., *Cyberactivism: Online Activism in Theory and Practice* pp. 165–184. London, UK: Routledge.

Gjoka, M., Sirivianos, M., Markopoulou, A., Yang, X. 2008. Poking Facebook™: Characterization of OSN applications. Proceedings of the first workshop on online social networks. Seattle, WA. August 18, 2008: ACM, 31–36.

Golder, S. A., Wilkinson, D., and Huberman, B. A. 2007. Rhythms of social interaction: Messaging within a massive online network. Third international conference on communities and technologies CT2007, East Lansing, MI. June 28–30, 2007.

Gould, S. J., and Weil, C. E. 1991. Gift-giving roles and gender self-concepts. *Sex roles,* 24:617–637.

Haddon, L., and Kim, S. D. 2007. Mobile phones and Web-based social networking—Emerging practices in Korea with Cyworld. *Journal of The Communications Network,* 6:5–12.

Hewitt, C. 2008. ORGs for scalable, robust, privacy-friendly client cloud computing. *IEEE Internet Computing,* 12:96–99.

Hinduja, S., and Patchin, J. W. 2008. Personal information of adolescents on the Internet: A quantitative content analysis of MySpace™. *Journal of Adolescence,* 31:125–146.

Jin-seo, C. 2008. Social networking embracing search engines, games. *The Korea Times* May 19, Retrieved May 27, 2008 from http://www.koreatimes.co.kr/www/news/nation/2008/2005/2133_24408.html.

Joinson, A. N. 2008. Looking at, looking up or keeping up with people? Motives and use of Facebook™. Proceeding of the twenty-sixth annual SIGCHI conference on human factors in computing systems, Florence, Italy. April 5–10, 2008, New York: ACM Press 1027–1036.

Kim, K.-H., and Yun, H. 2007. Cying for me, Cying for us: Relational dialectics in a Korean social network site. *Journal of Computer-Mediated Communication,* 13. Retrieved December 19, 2008 from: http://jcmc.indiana.edu/vol13/issue11/kim.yun.html.

Komaki, R. preprint. Do Web interfaces have politics? A case of mixi, a Japanese social network site.

Komter, A. 1997. Gift giving and the emotional significance of family and friends. *Journal of marriage and family,* 59:747–757.

Lampe, C., Ellison, N., and Steinfield, C. 2006. A Facebook™ in the crowd: Social searching vs. social browsing. Proceedings of the 2006 Twentieth Anniversary Conference on Computer Supported Cooperative Work, Banf, Canada. November 4–8, 2006, New York: ACM Press 167–170.

Lenhart, A., and Madden, M. 2007. Teens, privacy and online social networks: How teens manage their online identities and personal information in the age of MySpace™. *Pew Internet and American Life Project.* April 18, 2007. Retrieved May 2, 2008 from http://www.pewInternet.org/PPF/r/2211/report_display.asp.

Livingstone, S. 2008. Taking risky opportunities in youthful content creation: Teenagers' use of social networking sites for intimacy, privacy and self-expression. *New Media and Society,* 10:393–411.

MSNBC. 2007. MySpace™ deletes convicted sex offender profiles. *MSNBC News Services.* Retrieved July 30, 2008 from http://www.msnbc.msn.com/id/18686463/.

Parks, M. R. 2008. *Characterizing the Communicative Affordances of MySpace™: A Place for Friends or a Friendless Place?* Montreal, Canada: International Communication Association.

Pearson, E. 2007. Digital gifts: Participation and gift exchange in LiveJournal communities. *First Monday,* 12. Retrieved June 5, 2008 from http://firstmonday.org/issues/issue2012_2005/pearson/index.html.

Perkel, D. 2006. Copy and paste literacy: Literacy practices in the production of a MySpace™ profile. *Informal Learning and Digital Media.* Retrieved April 23, 2007 from http://www.dream.dk/uploads/files/perkel%2020Dan.pdf.

Raacke, J., and Bonds-Raacke, J. 2008. MySpace™ and Facebook™: Applying the uses and gratifications theory to exploring friend-networking sites. *Cyberpsychology and Behavior,* 11:169–174.

Rawsthorne, T. 2006. How pedophiles prey on MySpace™ children. *Mail Online.* Retrieved July 29, 2008 from http://www.dailymail.co.uk/femail/article-397026/How-paedophiles-prey-MySpace™-children.html.

Sherry, J. F. 1983. Gift giving in anthropological perspective. *Journal of Consumer Research,* 102, 157–168.

Thelwall, M. 2008a. Fk yea I swear: Cursing and gender in a corpus of MySpace™ pages. *Corpora,* 3:83–107.

Thelwall, M. 2008b. Social networks, gender and friending: An analysis of MySpace™ member profiles. *Journal of the American Society for Information Science and Technology,* 59:1321–1330.

Thelwall, M. 2008a. MySpace™ comments. *Online Information Review,* 33(1): 56–76.

Thelwall, M. 2006. Social network sites: Users and uses. In M. Zelkowitz, Ed., *Advances in Computers.* Amsterdam, Netherlands: Elsevier, pp. 18–73.

Tufekci, Z. 2008a. Can you see me now? Audience and disclosure regulation in online social network sites. *Bulletin of Science, Technology and Society,* 28:20–36.

Tufekci, Z. 2008b. Grooming, gossip, Facebook™ and MySpace™: What can we learn about these sites from those who won't assimilate? *Information, Communication and Society,* 11:544–564.

Walther, J., Van der Heide, B., Kim, S., Westerman, D., and Tong, S. T. 2008. The role of friends' appearance and behavior on evaluations of individuals on Facebook™: Are we known by the company we keep? *Human Communication Research,* 34:28–49.

Ybarra, M. L., and Mitchell, K. J. 2008. How risky are social networking sites? A comparison of places online where youth sexual solicitation and harassment occurs. *Pediatrics,* 1212:E350–E357.

Index

Printed and bound by CPI Group (UK) Ltd, Croydon, CR0 4YY

21/10/2024

01777103-0001